T0195307

Gene Jockeys

Gene Jockeys

Life Science and the Rise of Biotech Enterprise

NICOLAS RASMUSSEN

Johns Hopkins University Press

Baltimore

© 2014 Johns Hopkins University Press
All rights reserved. Published 2014
Printed in the United States of America on acid-free paper

2 4 6 8 9 7 5 3 1

Johns Hopkins University Press
2715 North Charles Street
Baltimore, Maryland 21218-4363
www.press.jhu.edu

Library of Congress Cataloging-in-Publication Data
Rasmussen, Nicolas, 1962– author.
Gene jockeys : life science and the rise of biotech
enterprise / Nicolas Rasmussen.
p. ; cm.
Includes bibliographical references and index.
ISBN 978-1-4214-1340-2 (hardcover : alk. paper) — ISBN 1-4214-1340-X
(hardcover : alk. paper) — ISBN 978-1-4214-1341-9 (electronic) —
ISBN 1-4214-1341-8 (electronic)
I. Title.
[DNLM: 1. Drug Discovery—history—United States. 2. Biotechnology—
history—United States. 3. Entrepreneurship—history—United States.
4. History, 20th Century—United States. 5. Technology,
Pharmaceutical—history—United States. QV 711 AA1]
RS380
338.4'76151—dc23   2013028965

A catalog record for this book is available from the British Library.

Illustrations by Cheri Cunningham and Nicolas Rasmussen

*Special discounts are available for bulk purchases of this book.
For more information, please contact Special Sales at
410-516-6936 or specialsales@press.jhu.edu.*

Johns Hopkins University Press uses environmentally friendly
book materials, including recycled text paper that is composed of at least
30 percent post-consumer waste, whenever possible.

CONTENTS

More than any other book I can imagine, this is the work of a lifetime in that I am drawing upon experiences since my teenage years, all specifically related to this work. I therefore owe debts to an unusually long series of people for enabling it. These include the biologists who took me into their laboratories and trained me in several fields, especially Michael Lew and fellow clinical microbiologists at the Dana-Farber, and Bruce Zetter and fellow angiogenesis researchers at Children's Hospital in Boston; Lucia Rothman-Denes and Peter Markiewicz and other prokaryotic molecular geneticists in the University of Chicago's Biophysics department; and all the outstanding biologists—too many to name—that I studied with at Stanford. They also include Tom Furlong and my business professors and my drug industry friends, and the fine people I worked with briefly in marketing services at a certain biotech company. They include my recent teachers and colleagues in public health and pharmaco-epidemiology, especially Lisa Bero, Simon Chapman, Ben Djulbegovic, and Adriane Fugh-Berman, from whom I have come to appreciate evidence-based medicine (EBM) both as a research program and an ethical perspective. Some readers may not appreciate the occasional "activist" stance in this book stemming from this perspective. I make no apology for holding that sick people should only be sold drugs proven equal to existing therapies for their condition; and that when patients are given drugs on an experimental basis in clinical trials, the results of this human experimentation should be available to all of medicine—no matter who pays the bills. Like EBM, this book is proscience, not antibusiness.

Of course I owe a great debt to a list of colleagues in history and philosophy of science too long even to begin, but one that particularly needs to include, for specific conversations that have informed this work over recent years, John Abraham, Mario Biagioli, Robert Bud, Bob Cook-Deegan, Ric Day, Jeremy Greene, Jim Griesemer, Paul Griffiths, Mary Hancock, Jon Harkness, Steve Hubicki, John

Krige, Toby Lazarowitz, Tim Lenoir, Harry Marks, Patrick McCray, Maureen O'Malley, Phil Mirowski, Peter Neushul, Scott Podolsky, Brad Sherman, Gabriela Soto-Laveaga, Karola Stotz, Bruno Strasser, and Peter Westwick. I also owe an enormous debt to all the scientists who took the time to interview with me: Julian Davies, Bruce Eisen, Ed Fritsch, Wally Gilbert, Alan Hall, Robert Kay, Glenn Larsen, Peter Lomedico, Tom Maniatis, Shige Nagata, Joseph Rosa, Peter Seeburg, John Shine, Daniel Vapnek, and Lydia Villa-Komaroff. I am very grateful to the Australian Research Council for funding the research, and to the Sydney Centre for the Foundations of Science for a sabbatical visit working on it. Finally, I must express my most grateful thanks to those who gave their time directly to this book: my partner, Jackie, for giving me space to write it, Andrea Gaede and Audra Wolfe for helping me research it and put it together, Cheri Cunningham for helping me realize the illustrations, Bill Summers and two anonymous referees for giving such constructive feedback on the manuscript, and Jackie Wehmueller for doing such a fine job in its publication.

Gene Jockeys

Biology's Day at the Races

"Gene jockeys"—as I learned as a biology graduate student at Stanford during the 1980s—were biologists working in the biotechnology firms that hired so many of my fellows at the time. Soon after, the term became the name of a lab computer program that found the correct position of DNA sequences. At a superficial level, the metaphor makes sense, because as a "gene jockey" a biologist spent most of his time moving DNA pieces from one place to another. But like many metaphors that catch on, this one is apt on a deeper level. Biotech in this early era was a festive contest closely akin to the premier horse racing circuit. Its venture capitalists, charismatic scientist-entrepreneurs, stock market investors, and of course its cloning races all have their close racecourse counterparts. And what does the metaphor say about the scientists, the main characters in this book? The typically ironic, self-deprecating but defiant usage reminds us that a gene jockey felt himself a smaller man in the eyes of academic counterparts, even if respected as an athlete.

Essentially, this book is about how the biotech arena emerged when molecular biology, one of the fastest-moving and most important areas of basic science in the twentieth century, met the business world during the 1970s and 1980s. It studies five of the first products to come from the new enterprise of genetic engineering. All these products were drugs, and all were previously known natural proteins, commercialized by cloning the human genes into microbes or other cells that could be grown in vats. Since pharmaceuticals were always the economic driver of the biotech sector, and furthermore since these same five drugs (and minor variants) today still account for an enormous share of the wealth said to be created by biotechnology, I think this study represents a fair basis for generalizations about the interactions of science and commerce in the biotechnology of the period.

Although it does not directly address the biotechnology industry of today, this book may interest business scholars and economists, among whom there rages debate about the extent to which biotech has lived up to the economic promises trumpeted in its early days, and the extent to which biotech's benefits have been increased or impaired by economic policies like the 1980 Bayh-Dole act that made it easier for US universities to patent publicly funded research.[1] In brief, the relevant implications of my story are that the early fruits of genetic engineering, low-hanging and ripened through decades of publicly financed biological research, may have been harvested earlier through the policy-stimulated, entrepreneurial rush for profit around 1980. However, with the general advance of molecular biology they would soon have been developed into medicines regardless—when the established drug firms learned how to clone. And for reasons just as firmly grounded in biology rather than economics, these findings imply that after this rich early harvest the field would necessarily yield less, and more grudgingly. There is no longer such a large backlog of fundamental knowledge ripe for "application" or, as current jargon puts it, "translation" as medical technology. Thus policies based on the premise that anything like this first bounty can ever recur are flawed. Moreover, I shall, on several grounds, suggest that in calculating the benefits from biotech enterprise in the era, there were also social costs, among which even the economically assessable ones are overlooked by economists, that need to be weighed along with the economic benefits as measured by sales revenues. For these mixed outcomes the policies that encouraged the rapid flourishing of recombinant drug development described here also need to be held accountable.[2]

But the focus in this book is on the science, and money's effect upon it and vice versa, rather than on the money per se. I will show that the development of the first-generation recombinant DNA drugs did not entail any great diversion or distortion of science. These projects were driven by biologists themselves, and in ways that grew out from traditional problems and values of the science(s) of molecular biology. Yet this is no "internalist" story of autonomous science, interacting only with society "outside" it to the extent that predestined discoveries are quickened or slowed (by funding, laws, etc.). Rather, this book's perspective is that the science of genetic engineering was—like all science—shaped by reciprocal interaction between the tradition-bound ideas and values of scientific fields, and the social institutions in which scientists participate, such as universities, funding agencies, philanthropies, medicine, and certain industries. When social conditions changed during the late Cold War period, particularly in the United States, this change affected these institutions and encouraged adaptation in the field of molecular biology. The already-perceived possibility of making drugs by genetically modifying microbes became more attractive, just at the same time that it was becoming more technically feasible. Rushing into the new socioeconomic niche that began (at first obscurely) opening before them, biologists actively reshaped institutions to facilitate their new activity: government funding agencies, their universities, regulatory law, intellectual property law, the stock markets, medicine, and also the popular understanding of life science. They also—more conservatively than often imagined—worked to adjust the values, intellectual focus, and reward systems in the scientific fields to which they still belonged. This book describes all these changes, while it also chronicles the remarkable, competitive, and at once secretive and spectacular science done by the gene jockeys who made the first generation of genetic engineering drugs.

Reconstructing this socially embedded story of the cloning races behind the drugs, while a novel and formidable project in itself, has also allowed the book to address some persistent arguments about whether close involvement with commerce is good or bad for science. When sociologist Emile Durkheim worried in the 1890s that industrial capitalism's huge success came at the expense of other social institutions, such as government and religion, he warned that science's quest for truth was similarly becoming subordinated to industrial ends.[3] The years around 1980 were experienced as the start of a similar period of crisis, this time deeply affecting science, as the postwar industrial preeminence of the United States waned, and the regime of generous state funding that had characterized early Cold War science policy began to falter. In the wider changes associated with Ronald Reagan, Margaret Thatcher, and the rise of neoliberal economic policy, social functions previously assigned to public institutions based on elite expertise began to be given over to "market forces," and the corporation again began growing in power at the expense of other institutions. In pharmaceuticals, regulation would be made less stringent on the grounds that rigorous testing before market approval unproductively delayed valuable innovations. In science, more university research would be funded by private investment, and thus produced for a commercial marketplace rather than scientific peers, on the grounds that the private sector generated useful goods more efficiently.

The move of academic molecular biologists into biotechnology projects—drug development being the most economically attractive—was immediately recognized as the vanguard of this shift in science, attracting hymns of praise and tempests of criticism from ideologues on opposing sides. Privately funded molecular biology would free a cornucopia of lifesaving new medicines from imprisonment in ivory towers, unleashing a storm of creative destruction in the pharmaceutical and agricultural sectors that might (incidentally) restore an American lead in high technology fast ebbing away to Japan. Alternatively, privately funded biology was gutting universities and diverting research away from areas of long-term benefit—discoveries about nature's basic laws with potential to spawn whole new industries, yet no immediate commercial product to offer. For example, the very techniques of molecular biology that enabled genetic engineering might never have been developed under this new regime; they came from government-funded academic research into bacteria and their obscure viruses over two decades, the 1950s and 1960s. So in effect, making science pay its way in the short run means killing the goose that lays the golden eggs.[4]

More recently, the critique has been refined, in ways at least a little less nostalgic for the Cold War, and the problematic features of market-oriented science better defined. As neoliberal state policy aims to produce public goods through private profit seeking, in science it encourages investment in publicly funded research and institutions by allowing private ownership of research outcomes. Market uptake of the resulting technology, not the judgement of impartial scientific experts, determines the success of research for those ascribing to these values—effectively leaving questions of scientific truth to the purported wisdom of nonscientist crowds. Leading critic Philip Mirowski, in particular, has charged that biotech was nothing but a business model imposed from outside biology, a deceptive investment scheme amounting simply to selling off publicly owned scientific assets to investors or big drug firms. Moreover, he argues, the model did not achieve its avowed goal of spurring biomedical innovation, producing instead a decline in the productivity of the life sciences by restricting cooperation between biologists and weakening peer-based quality control, by taking resources from the basic research in universities that generates fundamental value, and by hobbling efforts to bring new medical products to the clinic with excessive intellectual property baggage.[5]

In a number of ways the detailed story I tell fits with this broad brushstroke critique. However, I must differ on several points. Biotech was not a simple swindle, even if the public did pay a very high price for the development of preexisting public scientific resources (due, I argue, to a US intellectual property regime growing overgenerous in the 1980s). The work that scientists in the firms did went well beyond what biologists had already done in universities, and it certainly brought forward the medical use of the protein drugs I discuss by several years. Furthermore, and most importantly, we must remember that even if it was encouraged from outside, the biotech enterprise was one produced largely by biologists—and one with partial historical precedent in biology of the 1920s and 1930s. But there is one related aspect of this critique I must particularly endorse. Rather than an all-wise market guided by some superhuman invisible hand, there are visible and often heavy hands that construct and maintain the markets in public goods like science (which for ideological reasons are often pretended invisible). As I will show, in the rise of biotechnology many of the hands crafting the markets for genetic engineering drugs belonged to biologists, trying with their other hands (so to speak) to do science. There really was a moment when biologists could expect to do great science through the private sector, and hope at the same time for scientific acclaim and for material reward. This book tries to capture that moment—and the way the life sciences did begin to change, in what some might read as their tragic Faustian bargain.[6]

Structure, Sources, and Method

Conceived as a synoptic look at the whole realm of biotech from the late 1970s into the early 1990s, this book is a detailed study of the scientific and business projects leading to five of the first ten recombinant DNA drugs to be approved for medical use in the United States: human insulin, human growth hormone, alpha interferon, erythropoietin, and tissue plasminogen activator. Reflecting typical biotechnology in the era, these drugs are all first-generation recombinant products by a standard definition: naturally occurring proteins produced by recombinant DNA in such a way that they retain their native biological action.[7] The story of each captures in many ways a phase in the rapidly evolving field of genetic engineering. Because of its design around particular products and projects, the book focuses on the few firms and people most prominent in these five highly competitive drug development stories. The firms are paradigmatic and their activities fully worthy of detailed attention: Genentech, Biogen, Genetics Institute, and Amgen receive the most discussion because, scientifically, they did the most. (Granted, a selection of five different drugs among the first ten might have resulted in a somewhat different cast of main characters, but the overall picture would not differ much.)

In addition to a first chapter setting the stage, the chapters each describe one of these drug stories, cutting back and forth between the different scientists and firms at the forefront of their nearly simultaneous, rival efforts in molecular cloning—the process of isolating a desired human gene, splicing it into bacterial DNA, and moving that into a cultured cell where it will be reproduced (and, ideally, expressed). The different chapters, while ordered roughly chronologically, also overlap in time and are partly simultaneous, creating further narrative challenges that require references forward and back between chapters. The inevitable sacrifice in scope that results from this book's fivefold focus is justified not just by the detail enabled by selectivity, without which it would be impossible to capture the science adequately. It is further justified by the particular importance of the five featured drugs, both in quantitative financial terms and qualitatively, in terms of their impact on medical and public consciousness, political discourse, and the legal precedents that issued from the patent lawsuits surrounding them. These repercussions structured the subsequent actions of the scientists, businessmen, and firms involved in biotech and thus set the course for the next generation.[8]

In each of the chapters, I attend simultaneously to the lab science and to one of the broader dimensions along which the biotech firms and their scientist leaders reshaped their social contexts. In the preliminary chapter 1, where I describe the way basic research in molecular biology evolved in a Cold War context, almost inadvertently producing a toolbox for genetic engineering by the early 1970s, I also describe the way that prominent scientists then renegotiated the field's social contract in prominent political arenas so as to facilitate the shift to commercially funded gene splicing. In chapter 2, where I recount the three-way contest between Genentech and two commercially engaged academic labs to clone human insulin and produce it in bacteria, I also explore the way in which such first-generation drug development projects meshed with and carried forward existing intellectual and practical projects of fundamental research in molecular biology. I further discuss how the entrepreneurial biologists worked in public policy arenas to enlarge their overlapping zone of business and scientific opportunity. In chapter 3, I follow the continuing clash between a fledgling Genentech and a university biology group to clone and commercialise human growth hormone. Here I investigate the way in which the lab culture and scientific ethos of academic molecular biology of the 1970s was transplanted and grew in the new soil of the small biotech firm. In chapter 4, I describe the manyfold contest to clone interferon and therefore bring a much-anticipated cancer cure to market. Here, I look at how entrepreneurial academic biologists traded in the mixed markets for credit, scientific and financial, to encourage investment in their ventures. In chapter 5, where I reconstruct the typically fierce contest to clone and commercialise erythropoietin, I follow the cloning race into the courtroom, showing how this and related legal struggles engaged the higher federal courts in making de facto science policy with profound consequences for biotech. In chapter 6, where I describe competing efforts to bring tissue plasminogen activator to market as a heart attack drug, I also look at how the clinical trials required for regulatory purposes, together with the marketing needed to sell an approved drug, essentially required biotechs to become something very different—and very like the big drug firms.

Throughout these drug stories, I introduce interpretive tools to make sense of the wider implications that I am discussing, building on scholarship by a number of historians and social theorists including E. P. Thompson, Pierre Bourdieu, and Arjun Appadurai. A few are more original. I have done this unobtrusively (I hope) and avoided long theoretical excursions, both because I want the book to remain accessible to scientists and the general public, and because I believe in principle that the value of theory rests in its capacity to bring meaning and clarity to an empirically rich narrative.

I am acutely aware many of the events described here have been discussed before by others, in the enormous literature that has already accumulated around the fairly new phenomenon of biotech. However, nowhere can you find these five drug stories told in such detail, together in one place, and united in one historical analysis. And nowhere, I think, can you find the detail of the commercial molecular biology research related to academic life science at the time (even if not in quite the same detail), and also to the science involved in clinical testing. In addition to drawing on the rich secondary literature, I have relied on primary sources never before employed as a set by any one author. These include the scientific literature, of course, and also patents, which sometimes took so long to issue that they have only recently become publicly available; thousands of pieces of newspaper and trade press journalism, now much more accessible and systematically searchable through information technology; and also interviews with scientists by myself, by Stephen Hall and distilled in his outstanding book *Invisible Frontiers,* and (particularly for characters within the Genentech fold) by Sally Smith Hughes—made available through the University of California Library's Regional History collection in Berkeley.[9] These interviews have been invaluable especially in reconstructing the experienced, subjective side of the story.

Writing on corporate science presents unusual empirical challenges to a historian seeking harder documentary evidence, especially one pursuing his research without collaboration with the firms involved. To access proprietary information while still remaining impartial, I have gathered and drawn upon thousands of pages of corporate documents and testimony used as evidence in the intense intellectual property disputes surrounding all of the drugs discussed in this book. Although these court disputes, many of which only recently concluded, were an essential window on each drug's story and a sine qua non for its inclusion in this study, there is no reason to doubt the evidence so obtained or the resulting generalizations I derive about biotech. Firstly, it is hard to find a recombinant DNA drug of any consequence that was not the subject of intense intellectual property litigation. In this litigiousness the small biotechs differed from the big drug firms of the middle twentieth century, who considered patent infringement ungentlemanly and competed with one another via copycat or "me-too" drugs, compounds with essentially identical biological activity but different enough chemically to be patentable.[10] One may speculate that the litigiousness came from their upstart partners, the biologists and their little firms, who saw the courts as an arena to resolve disputes over scientific priority—or simply that the unsettled state of patent law around recombinant DNA invited litigation.

Second, only in the courtroom arena do some facts ever come to light. Evidence like lab notebooks, telephone calls, and private conversations, not typically preserved in academic archives or elsewhere, is assembled at great expense by large legal teams when millions are at stake, making past litigators research assistants to the historian. Thirdly, the adversarial context serves as a check—imperfect to be sure—on the soundness of the evidence, which enables a somewhat independent evaluation of statements by commercial actors made in print and in interviews for public consumption. And the historian, unlike his legal scholar colleagues who study the same cases, is free to bring his own interpretive framework to the evidence; he only has to interpret a court's decision as an historical event—as opposed to a precedent potentially useful in subsequent lawsuits (the hermeneutic mode unkindly called "precedent-mongering"). But like the courtroom, I understand history itself as an arena of clashing evidence-based narratives, constantly revised and revisioned through the diligence and critical ingenuity of adept historians. Thus I am confident that this book on the history of biotech can never be the final word.

CHAPTER ONE

Biology, Industry, and the Cold War

The ecstatic man steps from his wheelchair, no longer wracked with disease and smiling against the backdrop of a mushroom cloud. A grateful woman sobs with relief when her doctor, pronouncing her cancer diagnosis, also hands her a radioactive bracelet—the painless cure cheaper than a restaurant dinner thanks to "atomic energy." These images, one from a major magazine and the other from national radio, are typical of the flood sweeping American media in the wake of Hiroshima's and Nagasaki's annihilation in August 1945. The popular imagination fixated upon the idea of the atom's lifesaving power in the late 1940s, as if craving a redemptive silver lining in a thickening cloud of nuclear fear. This reaction thrust a previously obscure scientific field into an intense spotlight: biophysics. Biophysics, as its name implied both to scientists and the public, was the domain in which physics could contribute as much to medicine as it did to death and destruction. And at the time "molecular biology" was another, much less common name for biophysics. As we will see, what we now call molecular biology enjoyed a dramatic rise in the late 1940s and 1950s, buoyed by a rush of public attention and generous funding, and by a wave of scientific talent drawn to it—partly from physics.[1] But for about two decades before this sudden celebrity, molecular biology or biophysics was slowly emerging as a distinctive science. It is with this emergence that we naturally should begin.

Birth of a Science

As used by newspapers and the educated public in the 1940s, "biophysics" meant diverse things including sensory physiology and psychology, radiology, and subcellular biology as investigated with high-technology instruments like centrifuges. ("Molecular biology," by contrast, hardly appears in public discourse at all; there is scant evidence of the term's use even among scientists before the 1950s.)[2] Historians have traced the basic concept that all biological phenomena—embryonic development and mental activity and even life itself—were best studied in the simplest organisms and explained by ordinary physical laws, together with the biophysics label, to iconoclastic physiologist Jacques Loeb (famous in the early twentieth century for making sea urchins from eggs without sperm), and his students in the 1920s. The work of radiologists, biologists, and physicists on the physiological and hereditary effects of radiation upon living things had also fallen under biophysics by the 1930s. So too had the application of sophisticated technologies, many associated with physical chemistry, to biological problems: ultracentrifuges, electrophoresis apparatus, spectroscopes, isotopes and counters.[3]

This last kind of biophysics owes especially much to one man: Warren Weaver. As head of the Natural Sciences division of the Rockefeller Foundation, then the most generous private patron of science anywhere, from 1933 Weaver concentrated funding on particular areas of biology research that, he felt, promised to bring the rigor of physical science to bear on problems of health. Weaver especially supported research to adapt physical instruments and methods to investigate fundamental vital processes as they manifested themselves in mysterious submicroscopic entities like enzymes and viruses. Rockefeller-sponsored biophysical instruments and methods include what are now the most basic tools of molecular biology, the name Weaver himself coined in 1938 to talk about the new biology he vaguely envisioned. The scientists behind the tools operated at the margins of established scientific fields, and their research programs might well have perished during the Great Depression if not for Weaver's favor. With it they not only carried on, but some emerged from the war years with prominent university departments or research institutes of their own, effectively built with Rockefeller funding.[4]

Weaver was also very generous to research in two established life sciences already exemplifying the right (to him) rigor and technological orientation: endocrinology and genetics. As Weaver rhetorically asked the Rockefeller Trustees in his initial proposal to fund a "New Science of Man": "Can we unravel the tangled problem of the endocrine glands and develop a therapy for the whole hideous range of mental and physical disorders which result from glandular disturbance? Can we develop so sound and extensive a genetics that we can hope to breed in the future superior men?"[5] Although relatively new disciplines, these fields were already big and successful by the 1930s, thanks to the practical benefits they promised and partly delivered.

Classical genetics had risen to prominence in the United States in the 1920s.

Although the main practical benefits of the field, like hybrid corn, came from plant geneticists, the greatest fame belonged to Thomas Hunt Morgan and his school of fruit fly geneticists. These researchers invented the methods by which genes, hypothetical units of heredity known only by the abnormalities they produced when defective (e.g., bar eye, curly wing), could be mapped on the chromosomes according to the frequency with which two different traits were inherited together in breeding experiments. Because genetics promised agricultural improvements, the field was relatively well funded by legislatures, and even in the Depression geneticists were able to find jobs at public universities in the United States. The field also owed its continued growth in the interwar period to the eugenics movement, popular ambitions to improve the human race that American geneticists drew upon and did not disown, throughout the racist and isolationist 1920s and well into the 1930s. Witness, for example, the feature in November 1934's *Popular Science* on the Morgan lab's work mapping fruit fly genes, which was headlined "Sensational Study of Heredity May Produce New Race of Men." Some historians have seen the human genome project, that recent milestone of molecular genetics, as a realization of the eugenic agenda behind both genetics and Weaver's envisioned new science of man.[6]

As for endocrinology, the science that could correct the "glandular disturbances" underlying so many medical and social ills, there was no scientific field more fashionable during the 1930s. Surgical experiments on animals during the late nineteenth century had pointed to certain glands as controlling physiological processes through circulating chemical signals (or "hormones"). The idea that health and illness were largely a question of glands swept medicine early in the twentieth century. Hormones quickly drew the interest of biochemists seeking to isolate the messenger compounds, and of drug companies seeking to sell them as medicines. Around 1900 adrenaline (norepinephrine) was the first hormone purified by biochemists, and it quickly became a blockbuster drug for several medical needs: raising blood pressure or preventing surgical bleeding when injected, and relieving congestion in spray form. Pure thyroid hormone (thyroxine) was isolated and marketed for weight loss in the late 'teens, insulin for diabetes in the 1920s, and a range of both estrogens and androgens in the 1930s for "feminine troubles" and flagging virility in middle aged men, respectively. The hybrid research field of endocrinology, in which medicine and chemistry overlapped, was highly competitive and well financed. The fame to be won by the scientist who first isolated the pure hormone from a particular gland, and the profits to be made by the drug firm with exclusive rights to that hormone, spurred what one endocrinologist in 1929 called a "gold rush" of collaboration between corporations and academic life scientists (who were following in the footsteps of chemist colleagues with industrial links). At the same time these commercial interests reduced cooperation and communication among scientists in the field, which was already more competitive than genetics, a publicly funded area with a culture of openly sharing fly strains and discoveries throughout a global community of researchers.[7]

These endocrinological collaborations, where a drug firm provided a scientist with resources and technical assistance with the industrial-scale demands of hormone research, and the scientist provided the firm exclusive access to a new hormone drug, set off a wave of patenting among university life scientists between the wars. For example, shortly after Edward Doisy isolated an estrogen in crystalline purity in 1930, he licensed his preparation patents to the Parke-Davis drug firm, which in turn provided Doisy with extensive research support, thus helping him extract other estrogens from tons of pig ovaries. Variations on this arrangement can be found in the collaborations behind the development of testosterone, cortisone, Vitamin D, and the anemia drug cobalamin (now Vitamin B12). The university scientists involved would typically receive royalties from licensing their patents exclusively to the drug firm partner. Apart from sourcing new products, another key way in which industrial partners could benefit was by gaining prestige and marketing advantage through another form of intellectual property, the trademark, as when Lilly obtained exclusive rights to sell insulin under Banting and Best's name for the hormone, Iletin, and when Parke-Davis marketed Doisy's estrogen as Theelin. Some endocrinologists, like Doisy, even allowed their own names to be used in drug advertising, for example a full-page 1931 ad for Theelin declaring that "the credit for the isolation of this hormone belongs to Dr. E. A. Doisy, professor at St. Louis University," who "entrusted" its manufacture exclusively to Parke-Davis. Some universities banned such use of their name, suggesting both that the inclination to do this was not rare, and that it was considered problematic for a university's reputation. The patenting wave did generate controversy, both over the sanctity of science and the monopolies over medicines that it led to, but not enough to prevent Nobel Prizes for many participants (including Doisy, albeit for another accomplishment).[8]

The hormones that became the era's blockbuster pharmaceuticals were either extracted from slaughterhouse organs by scaled-up versions of the isolation procedures used by the scientists who discovered them, or else reproduced from biological starting materials by chemical modification. Others were discovered but resisted commercialization, either because they were too hard to manufacture in useful quantities or because the animal hormones did not work in humans.

Endocrinology's success in bringing the fruits of biochemistry to human medicine impressed Weaver enough to throw more money at it, even though it was already financed by the drug industry. It would take nearly fifty more years for Weaver's pet project, molecular biology, to spawn a gold rush of its own, but eventually it did. And as we will see, when that gold rush came, the molecular biologists followed in the endocrinologists' footsteps in many ways, not least of which was that the targets most avidly pursued were hormones.

Molecular Biology in the Atomic Age

After Hiroshima biophysics enjoyed sudden fame and fortune, as the field promising to reap wholesome benefits from the same science, physics, that had so magnified the forces of death. In the late 1940s this expanding field still encompassed life sciences related to light and radiation (for instance, research on photosynthesis and radiation genetics) as well as nearly anything involving sophisticated instrumentation or the use of atomic isotopes as tracers. Soon the term as popularly employed included a new area of research, the study of genes in the simplest living things, bacteria and bacterial viruses—microbial or molecular genetics. This assortment of scientific projects, united largely by their focus on the simplest units of life, and their loose connection with the theories and tools of physics (and also their lack of connection with mainstream biochemistry as practiced in the 1940s), was the kind of biophysics that developed into what we now call molecular biology. Other parts of what was then biophysics now reside in different fields.[9]

Because of the cultural resonances of biophysics, its various practitioners benefited enormously from a blend of political will, magical thinking, and scientific opportunity in the early atomic age—particularly in the United States. Already in 1945 the University of Chicago, "reluctant to disband the team of brilliant scientists who produced the first nuclear chain reaction" in the Manhattan Project (as publicity put it), organized a $12 million set of Research Institutes. Dedicated to researching peaceful uses of the atom these Institutes, including the Institute for Radiobiology and Biophysics, were to be funded entirely by donations and industry subscriptions. Billboard slogans like "Turn atomic power on cancer" were part of the successful initial campaign that supported the work of former bomb-makers like James Franck on photosynthesis, and Leo Szilard, who studied mutation in bacteria. At the University of Michigan Robley Williams, a struggling physicist-turned-biologist unwelcome by the biology department, suddenly found himself in high demand when in 1948 the university established a special fund called the "Phoenix Project," devoted to making

"the potentialities of atomic energy . . . a beneficent influence on the life of man."
The year 1949 saw Williams speaking at trustee luncheons and alumni fund-
raising dinners, and posing with his electron microscope in the alumni newsletter, for which he was rewarded when the university's Regents voted to establish
a department of biophysics. Encouraging many universities along these lines,
in 1947 Congress imposed an earmarked life science budget on the reluctant
leadership of the new Atomic Energy Commission (AEC), on the grounds that it
was unseemly for the United States to be spending so much money on the atom
solely (as Representative Everett Dirksen, R-IL, put it) "to perfect an instrumen-
tality for killing people." To meet this requirement AEC began pumping out
radioisotopes for medicine and biology, and funding life science research. Many
molecular biologists and other academic biophysicists received generous AEC
grants. Some, like University of California, Los Angeles radiation biologist Stafford Warren, previously medical chief at the Alamogordo (Trinity) and Bikini
atomic tests, built entire biophysics departments with AEC backing.[10]

That the American public was now virtually force-feeding biophysics was
noted by established leaders in the field, like Weaver favorites neuroscientist
Francis Schmitt and virologist Wendell Stanley. In talks on the radio and other
popular venues in the late 1940s, both took the opportunity to supplement their
already plentiful funding by representing their research on the physical basis of
life as a wholesome alternative to atomic physics. The postatomic enthusiasms
these biophysicists were tapping underwrote the institutions where molecular
biology blossomed in the 1950s, like the life science department Schmitt built
at Massachusetts Institute of Technology (MIT) and Stanley's Biochemistry and
Virus Lab (later, Molecular Biology and Virus Lab) at the University of California, Berkeley. Even Linus Pauling, who personally disliked the term "biophysics," employed it amply if a little cynically in his successful late-1940s applications for the massive Rockefeller Foundation funding that made Caltech into
Weaver's main postwar showpiece and powerhouse of the new biology. Yet this
enthusiasm reflected far more than vulgar fashion. Within the scientific community in 1946, biophysics suddenly became just as fashionable, as the spate
of Nobel Prizes that went to Americans for biophysical work that year suggests.
Wendell Stanley shared the chemistry prize for crystallizing Tobacco Mosaic
Virus in 1936 (and thus showing that these self-reproducing "giant molecules"
were ordinary chemicals), and geneticist Herman Muller won the life science
prize for showing in 1926 that radiation caused mutations, thus opening the way
to genetic engineering—and also to the study of the atom's genetic perils, his
new political cause.[11]

Responding to the flood of new resources, talent, and prestige, the science now thought of as molecular biology was thriving by the late 1940s, at MIT and Berkeley and Caltech and at other American universities too (Johns Hopkins, Pittsburgh, and Yale all offer further examples of new biophysics departments). One nonacademic institution also deserves special mention: the Cold Spring Harbor Laboratories (CSHL) on Long Island. Established as a marine station at the turn of the century and then augmented with genetics and eugenics research facilities, CSHL became a leading site for biophysics when radiation geneticist Milislav Demerec was appointed director in 1941. Through the work of both permanent scientific staff and the many visitors who congregated there for summer research and short courses, CSHL became a biophysics hub in the 1940s and 1950s, and is especially famous for "phage course" in the genetics of bacteriophage (bacterial viruses) that former physicist Max Delbrück offered there from 1945.[12] The new biology's growth spurt slowed after about five years, when an economic downturn and the Korean War pinched federal funding. But the science had been set in motion and the institutions well established so that when government generosity returned and growth resumed in the mid-1950s, in a new funding regime and political climate, biophysics and its increasingly popular semisynonym "molecular biology" flourished still more.

Problems and Products of a New Science

The intellectual preoccupations of biophysicists in the late 1940s and 1950s had to do with the mechanisms of life below the scale that could be directly observed by eye and microscope. How did giant "macromolecules" perform life's amazing functions, like metabolism and motion? The premise that the answers required understanding how these substances—especially proteins—made up minute machinery in the living cell put a premium on technology to separate and visualize them, and monitor their action, via ultracentrifuges and electron microscopes and the various other biophysical instruments and methods. The effects of radiation on living cells was also a core problem of the field in this early period, with the genetic side of this question (as "DNA repair") remaining a part of molecular biology through the 1960s. And, as so many histories of molecular biology have stressed, biophysics' preoccupations included (but were not limited to) the mysterious physical basis of the gene: how can a molecule make a copy of itself, and also make the enzymes and other substances that genes were known to contribute to the body? This—the twin problems of heterocatalysis and autocatalysis, as theorists put it between the wars—was the focus of postwar molecular genetics. The 1953 Watson-Crick double-helix model of DNA only pointed to a solution for the first part of the problem, autocatalysis, by showing that the sequence of bases on each strand of a DNA molecule could specify the other. How this duplication of the genetic material actually occurred in cells was yet to be determined. And the riddle of heterocatalysis, how the genes made something other than themselves, remained unsolved.[13]

Well before the double helix, biochemists had been making some progress on this latter problem, although they saw it as a question of characterizing the enzymes and other factors involved in protein synthesis. In the 1930s cytologist Albert Claude at the Rockefeller Institute developed methods to homogenize living tissue (typically, liver), and to separate out fractions of the soup by centrifuging, that allowed the protein-producing reactions of the cell to be reconstituted in a test tube. By the 1950s other researchers, particularly biochemist Paul Zamencnik's group at Harvard medical school, built on this cell-free approach to identify the components required for protein synthesis, now traced through radioactively labelled amino acid building blocks. One essential component, particles called microsomes by Claude (and later renamed ribosomes) seemed to determine which particular proteins were made. Together with the microsomes and the amino acids themselves and some other simple chemicals, Zamecnik identified a long list of enzymes involved in protein synthesis, one for each of the 20 common amino acids, and also a soluble form of a sugar polymer related to DNA: RNA.[14] The biochemists were gradually putting together all the pieces of the puzzle, except where genes fit in.

In the mid-1950s, biochemists similarly began studying DNA synthesis (that is, chromosome replication) in the test tube using much the same approach that they were simultaneously turning on protein synthesis. Around 1955, Arthur Kornberg isolated an enzyme from homogenized *E. coli* bacteria that, given a single-stranded DNA template and nucleotide building blocks (as well as a short "primer" strand of DNA complementary to the template), made more DNA based on the template. Apart from this DNA-synthesizing or "polymerase" activity, the enzyme also had two other, separable activities, the capacity to degrade DNA from either of its chemically distinct ends. Although Kornberg's DNA Polymerase I, as it came to be called, turned out to be a DNA repair enzyme rather than the main enzyme devoted to chromosome replication, its discovery gave biologists a key instrument to manipulate DNA strands in the test tube by chewing back single stranded DNA or making it double stranded, and also making copies from any desired template. This work was very quickly recognized with a 1959 Nobel Prize and a leadership position at Stanford, which Kornberg helped make into a molecular biology powerhouse in the 1960s. And although it took a slower route from research subject to routine experimental tool, by the late 1960s molecular biologists also had at their disposal DNA ligase, an enzyme extracted from bacterial viruses that was useful for permanently (re)joining broken DNA strands.[15]

The enzyme toolbox was growing quite diverse by the mid-1960s, with large batches being routinely prepared from bacteria and viruses by technicians in the ever-larger laboratory groups of molecular geneticists—most of whom had little or no formal training in enzymology. Rather than making everything they needed for themselves, each lab group tended to specialize in producing only a few enzymes using time-honored recipes. Following the communalist traditions of the fruit fly geneticists (traceable, perhaps, to the key role Delbrück and Caltech played in the formation of molecular genetics), they shared small samples of enzymes and new and useful strains of microbes with almost anyone who asked. For larger quantities they traded, in the manner of craftsmen in traditional gift-based cultures, with different lab groups producing the other enzymes they needed. By analogy with the trading languages that historically connected peoples along far-flung maritime routes, one might say that in its maturing stages molecular biology depended on a "pidgin enzymology": a shared set of techniques and products that was somewhat crude compared with the enzymology practiced by biochemist specialists, but adequate to support the community's practices. In the later 1960s, however, some molecular geneticists were beginning to distinguish themselves—and their field—by discovering new enzymes without the help of card-carrying biochemists (although at this stage the gulf between biochemistry and molecular biology was shrinking).[16]

By the middle 1950s, biochemists studying proteins were becoming convinced that the particular order of the 20 amino acids making up the chain determined the way that protein would fold in the cell, and therefore the biochemical activities that protein would exhibit. Pauling's 1951 prediction of the alpha helix structure—confirmed with direct evidence six years later—was based purely on careful models accurately reflecting bond angles and physical chemistry constraints, and stands out as one milestone. So too the experimental work of NIH biochemist Christian Anfinsen, on the folding and refolding of the small enzyme Ribonuclease A: thermodynamics alone seemed to guide a protein with a given amino acid sequence to its final, correct, and functional shape. Thus amino acid sequence determined the function of proteins. And as already known since 1949, from work on sickle cell hemoglobin, genetic mutations altered amino acid sequence.[17]

It was here, at the intersection of sequence, structure, and function, that biochemistry met genetics. In 1953, Francis Crick and James Watson had specu- lated that the sequences of nucleotide bases in a gene's DNA might determine the sequence of amino acids in the corresponding protein. Since 1954, they had been trying with a circle of theoretically inclined friends to "crack" the prob- lem of how the DNA sequence "encodes" protein sequence through abstract cryptographic methods. Crick brought all of this together in a 1957 talk that articulated what would become molecular genetics' unifying theory. Called the "central dogma of molecular biology" (a label Crick in particular preferred over biophysics), the theory held that genetic information resides in the DNA sequence and flows to amino acid sequence in protein, and only in that direc- tion. Between DNA and protein stood an obscure intermediate process possibly involving RNA. Crick called this unidirectional flow a dogma in recognition that the concept could not yet be proven (indeed, the idea that DNA sequence infor- mation corresponds one-for-one with protein sequence was not proven until 1964, and the idea that genetic information flowed only one way between DNA and RNA proved wrong).[18]

The looser speculations of Crick and his circle had included talk of RNA adaptors between the DNA code and amino acids as well as short-lived, vaguely conceived messengers that relayed environmental signals to the DNA. Bacterial geneticists further developed the concept of a gene-specific, short-lived form of "informational RNA" as they learned experimental tricks to monitor specific gene expression events. In 1961 bacteriophage researchers substantiated the con- cept by showing that freshly infected bacteria made new RNA and then new virus proteins using their old ribosomes. Although half a dozen research groups were hot on the trail of this transient informational RNA, the entity we now rec- ognize as messenger RNA only became a concrete, manipulable substance in the hands of the biochemists diligently trying to reconstitute protein synthesis in the test tube. That same year two groups simultaneously found that the addi- tion of a synthetic RNA consisting of a long string of just one nucleotide base to their cell-free systems (derived from bacteria, not liver) produced a protein con- taining just one repeating amino acid. The biochemists had not only proven the existence of a "messenger code" linking RNA and amino acids and displaced the still-dominant notion of the ribosome particle as a specific template. They also found a way to decipher Crick's "code" by creating artificial messenger RNAs with known sequences, then analysing the proteins made in their test tubes.[19] The power of this technique was astonishing: the code for all 20 amino acids and 64 coding triplets was entirely deciphered within a few more years.

Through the 1950s, while the problem of how genes related to proteins yielded to impressive gradual progress, the longstanding puzzle of regulation stood virtually untouched. Classical genetics had posed but never addressed the central problem: how every cell in a complex organism manages to activate only those few genes imparting its special identity (as liver cell, neuron, or whatever), even though genetically identical to every other cell in the body. Molecular geneticists framed this as a problem of gene expression, or what makes a gene produce or not produce the protein it specifies. Their workhorses, the bacteriophages, offered a simple and defined set of gene expression events, in the virus infection life cycle. After 1945 they were also able to apply genetic methods to bacteria themselves, enabled by the sex-like "conjugation" process of gene transfer discovered by Joshua Lederberg. Research on the problem of gene expression and its regulation in these simplest of organisms was from its earliest days said to model not just gene action and the normal development of the embryo, but also the fearful abnormal development, that is cancer, providing molecular genetics with a compelling claim to medical relevance and medical research funding. In the domain of regulation a breakthrough, one truly deserving of this overused term, came from a small enclave of bacteriologists at the Pasteur Institute, who had taken that institution's tradition of research on bacteria and their adaptation to growth conditions in the new, internationally fashionable molecular genetic direction. There André Lwoff, Jacques Monod, and Francois Jacob were investigating the regulation of the enzyme, beta-galactosidase (or B-gal), in *E. coli*. These bacteria normally produce B-gal, which digests the milk sugar lactose, only in the presence of that sugar. Lactose "provoked" or "induced" the enzyme; the absence of lactose or a similar sugar suppressed production of the enzyme.[20]

Like all geneticists, the Pasteurians collected mutants to study the phenomenon: some mutant strains were unable to break down the sugar, because of a defective B-gal enzyme; others had a different defective gene for an enzyme needed to transport lactose into the bacterial cell. Through Lederberg they found another mutant strain in which the bacteria produced lactose-digesting enzymes constantly or constitutively—whether or not lactose was present. Using Lederberg's new conjugation-based genetic mapping techniques, they discovered that the constitutives were of two distinct types, mutated at two different spots on the chromosome: one near the inducible enzyme genes, and one remote from them. A defect at either of these two sites appeared to make whatever turned the enzyme genes on or off insensitive to the nutritional environment. Putting all of these genetic data together, the Pasteur group proposed a startling theory in 1960–61. First, they hypothesized the existence of a special "regulatory" class of gene, whose function was to switch the ordinary genes making proteins—now dubbed structural genes—on or off. Second, they proposed that their regulatory and structural genes together constituted a single self-regulating unit they named the "*lac* operon." In this theory, the structural genes were not usually making enzymes because a nearby regulatory site on the chromosome, named the operator, was bound to a special molecule they called a repressor. But in the presence of lactose or similar sugar, the repressor molecule—which turned out to be a protein—bound to the sugar and thus released its grip on the operator site, allowing the sugar-metabolizing genes to be activated. When the bacteria had consumed all available lactose, the repressor would again bind to the operator site on the chromosome, shutting down the structural genes in the operon once more. Defects in the operator site and in the gene for the repressor protein corresponded to the two types of constitutive mutant. Key aspects of this automatic feedback control concept came from Szilard, who was visiting the Pasteur and saw in gene control deep similarities with cybernetic technology of the day, such as in guidance systems.[21]

After 1961 this operon theory of gene regulation quickly became not just a useful example but a foundational concept, along with Crick's central dogma, for the discipline of molecular biology (or at least its increasingly dominant part, molecular genetics). If we are to follow Thomas Kuhn, whose theory of science articulated and was adopted as the official philosophy for American scientists at about the same time (see below), the operon model became a key element of the "paradigm" defining that discipline and the basis on which, through extension of the paradigm to new phenomena, it evolved into a "normal science." A handful of other bacterial gene clusters were found to be self-regulating in similar ways. Some were arranged, like the *lac* operon, so that the default position without the triggering chemical was "off." In others the default position was "on," as in the *trp* operon controlling expression of the enzymes needed to produce the amino acid tryptophan (when present in sufficient quantities, the amino acid interacts with the repressor protein to block expression of the genes that make more of it). In either case, the core concept was that regulatory genes controlled the expression of suites of other genes, often via environmental signals indicating when they should be "transcribed" into messenger RNA. The rapidly expanding menagerie of gene regulation systems found in bacteria and their viruses, illuminated by normal science in molecular biology, were all conceivable as variations on this cybernetic operon theme. The obvious next frontier was to extend the paradigm by moving, in Jacques Monod's famous dictum, from *E. coli* to elephant: to discover the genetic control mechanisms of higher organisms like flies and mice and humans—the eukaryotes.[22]

For its first two flourishing postwar decades the new science of molecular genetics was to its practitioners a basic science directed at uncovering life's fundamental workings, and perhaps also the causes of cancer. But like the neighboring field of biochemistry, which gave rise to many useful new products (like enzymes) and techniques (like chromatographic methods allowing both discovery and production of macromolecules) in the pursuit of knowledge, molecular genetics' ways of discovering truths of nature were also ways of generating practically useful new things—some of them living. For example the Pasteur group showed that the *lac* repressor locus made a diffusible product (a protein) and that the operator locus was its binding site on the chromosome, by creating bacterial strains with extra copies of these genetic elements using Lederberg's gene transfer method. Strains carrying various versions of the *lac* operon would later be crucial in genetic engineering. George Beadle, who shared the 1958 life science Nobel with his biochemist partner Edward Tatum (and also Lederberg), had earlier proven that genes encode individual proteins by creating mutant strains in the bread mold *Neurospora*, using X-rays, that lacked particular enzymes needed to make particular amino acids. Apart from revealing "the very elements of heredity" (as the Nobel presentation put it), this collection of mutants helped biochemists trace the metabolic pathways involved in the biological synthesis of the amino acids, and likewise constituted an instrument to measure the quantity of those amino acids in foods—and was used this way by manufacturers during the World War II. Leo Szilard and Aaron Novick developed the chemostat, a device to culture bacteria under constant conditions, in order to measure mutation rates caused by ultraviolet light and chemical mutagens—which enabled them to search for chemicals that reduced mutation rates, and thus might become antiradiation drugs. Ananda Chakrabarty eventually used this device in creating his new and patentable strains of pollution-consuming bacteria at General Electric.[23] So molecular biology was from the beginning creating new technologies and entities with potential medical and industrial (and military, conceivably) utility, even in its most fundamental discoveries, and even before it turned its attention from microbes to higher organisms in the 1960s. In the early postwar years, the potential utility of their work seemed a secondary concern to many practitioners of basic molecular biology—perhaps worthy of brief speculation in the introductory sections of grant applications, perhaps unworthy of mention at all. This attitude toward utility would change. But before we sketch the further development of molecular biology in the later 1960s, a short digression is required on the cultural meaning and politics of basic research in molecular biology during the formative Cold War years.

A Normal Science for the Cold War

The year 1962, when the Cuban Missile Crisis brought Cold War tensions to their extreme, was a banner year for molecular biology—viewed both as a cluster of research fields, and a label or brand. Watson, Crick, and Maurice Wilkins received the life science Nobel Prize for the double helix model of DNA, while John Kendrew and Max Perutz won the chemistry Nobel for their crystallographic work on hemoglobin protein. In Cambridge, England, where all this great work took place, the institution that had housed it was reinvented as a new Laboratory of Molecular Biology (or LMB). Crick's new division was called "molecular genetics"; reflecting the broad construal of molecular biology at the time, the other two divisions of the LMB were "structural studies" (for X-ray crystallography) and "protein chemistry." At other institutions, this latter field remained in biochemistry departments, as the discipline of biochemistry increasingly assimilated the new physical chemistry thinking and methodology promoted by Linus Pauling at the expense of the older constellation of fields known as "biophysics." Biophysical cytology, the investigation of cell structure with the electron microscope, split off as "cell biology" at about the same time. As the old, broader biophysics broke up and the molecular biology cluster of fields within it became more of a distinct discipline, molecular genetics moved to the center. The early contents of the *Journal of Molecular Biology* (founded 1959) reflects this narrowing to the study of gene expression, mainly in bacteria and bacteriophages, and macromolecule structure. By the end of 1962, Western European leaders of the field were laying the foundations of a new European Molecular Biology Organization (EMBO), to make support of this science a scientific pillar of the new European Economic Community binding the core NATO nations. And 1962 saw the NSF's life science division scrambling for creative new ways to pump money into molecular biology and thus steal limelight, even briefly, from the NIH as the premier US patron of the era's "glamour field."[24]

Coincidentally, 1962 was also the year in which historian and philosopher Thomas Kuhn published his enormously influential book, *The Structure of Scientific Revolutions,* which both described and justified the new social role science came to play in the early Cold War years. Kuhn argued that fields became true sciences when they stopped debating internally over fundamental premises and methods, and instead reached consensus on a fundamental set of beliefs and practices that he called the "paradigm." Once a research field found its paradigm, it entered into a phase of "normal science" that Kuhn likened to routine "puzzle-solving," in which scientists spent their time indefinitely extending and refining that paradigm (until, that is, a rare revolutionary like Darwin or Copernicus breaks it). Kuhn was building philosophically on a theory of sociologist Robert Merton, who had argued since the late 1930s that science flourished only, and could only ever flourish, when the scientific community functioned autonomously in imposing its own values and reward systems. This autonomy stood in contrast to the sort of political interference associated with the "totalitarian" regimes of Hitler, who rejected quantum mechanics as Jewish physics and appointed hacks believing in Atlantis to head anthropology institutes, and Stalin, who sent geneticists to labor camps and from about 1940 (to about 1960) made the Lamarckian ideas of agronomist Trofim Lysenko into official Soviet doctrine. On this view, only in the Free World could science truly thrive, an ideological stance that provided the United States with valuable international credibility in the early Cold War. Thus as a sign of its superiority, the United States boasted and therefore needed to support outstanding free inquiry, especially in natural science. Incidentally, the same stance provided cosmopolitan scientists in the United States a shield against political interference by Senator Joe McCarthy and his witch-hunters. As a price of their protection from politics, though, American scientists were expected to remain strictly apolitical or suffer consequences, as we will see.[25]

Kuhn also built on a new postwar sociological reality at American universities, where science was funded with unprecedented generosity by the state. During World War II, academic science and scientists were harnessed for military research on a large scale, for the massive atom bomb effort and also for a host of other projects both famous—like radar and penicillin—and secret, like biological warfare. Consequently, a great many scientists grew accustomed to government support. The biomedical arm of OSRD, the premier but by no means only wartime science agency, alone funded 5,500 life scientists under 600 research contracts, most of which were based at universities. The head of OSRD, Roosevelt's science chief Vannevar Bush, attracted postwar limelight with his appeal that the Federal government should continue funding academic research, particularly curiosity-driven "basic" research (a term largely displacing the older, impractical-sounding "pure" research). Such science would provide the basis or foundation for industrially-oriented "applied" researchers to build the next generation of technologies, the thinking went, assuring long term national security and wealth. After much debate, the National Science Foundation (NSF) was finally established in 1950, but it took until the late 1950s (following Russia's 1957 Sputnik triumph) for NSF to become a major funder of biology. However, NSF was never the largest sponsor of academic biology, because it was always dwarfed by the ever-expanding extramural research budget of the National Institutes of Health (NIH). Mission-oriented defense agencies like the AEC and the Office of Naval Research also funded academic science extensively during the 1950s, including biology, partly to advance technological agendas but dimly perceived by scientists and partly to cultivate goodwill. Together all this government funding produced what historian Roger Geiger has called a "hypertrophy of disinterested research" in American universities, especially after Sputnik. Kuhn's concept of "normal science" responded to the situation by suggesting that the research program of a science emerged strictly from the internal logic of that science's paradigm, and need not serve any other purpose but to solve puzzles. Therefore only full members of a scientific discipline were equipped to judge what research directions were most promising (as well who should get credit, as Merton had argued).[26]

Kuhn thus rationalized the public financing of scientific fields through small numbers of peer reviewers representing the top scientists in the field (especially those who had recently won grants), the method already relied on by NIH and NSF, and which in the post-Sputnik era became "normal." This concentration of the purse-string power in the hands of the scientific elite had the effect of making the credit-rich richer, and funnelling the lion's share of public funds to a handful of top universities, where entrepreneurial scientific leaders managed the large lab groups best equipped to win the biggest grants. This wealth distribution side effect produced political friction both at the time and ever since (as did that of the main alternative to project-based, peer reviewed funding, block grants to states or institutions). Peer review had the additional worrying scientific side effect of narrowing research by reinforcing the mainstream thinking defined by the elite peer review committees. But by making group-think seem a necessary and natural consequence of any scientific paradigm, Kuhn eased these concerns. Kuhn's theory was thus congenial to established fields and elite institutions, and became a linchpin of the ideology of basic research that historians see as dominating American science in the early Cold War years.[27] Whatever their funding agencies may have thought or told Congress about ultimate practical payoffs, amongst themselves academic scientists generally justified their work as contributions to basic knowledge in their discipline. Most were true believers in this ideology of disinterested, curiosity-driven basic research as a socially beneficial calling. Practical realization of the benefit was for someone else to deal with, according to the accepted division of labor.

In a context where even physicists developing the maser (microwave laser) under classified military contracts could view themselves as merely advancing basic knowledge for its intrinsic value, still more so could molecular biologists, whose work's contribution to the Cold War was less direct, take on the mantle of apolitical scientists' scientists, pursuing science only for science's sake. As Stanley put it in 1950, when representing academic staff at the University of California opposed to its anti-Communist Loyalty Oath, "The Faculty . . . always will vigorously oppose any system, Communist or other, which does not permit freedom of inquiry, of opinion and of teaching." For Stanley, a politically imposed loyalty oath for scientists whose engagement with the free exchange of knowledge already demonstrated their loyalty, and fulfilled their patriotic duty, was redundant. (Indeed, it undermined the emerging function of the nation's prestigious universities in the Cold War; as a high-level 1953 report on the federal government's propaganda priorities noted, it was essential to contradict world opinion that Americans are belligerent philistines and "that the spirit of the American government is not very different from that of the notorious witch-hunters within it.") However, Stanley continued reassuringly, Communists and "fellow travellers" had "no more chance of getting on the faculty here than Mickey Mouse." He thus strayed close to self-contradiction in a way typical of liberal intellectuals at the time, such as the many who condoned blacklisting of artists and musicians. Certainly, while by no means alone in his vocal opposition to nuclear testing in the 1950s, Linus Pauling was a lonely figure compared with the atomic scientists who had clamoured for disarmament before the Cold War set in. And he suffered for it, for example by being blackballed for university jobs and having his passport revoked (and when he received the 1962 Nobel Peace Prize, *Life Magazine* called it a "weird insult from the Norwegians"). Political engagement became suspect intellectually as well as politically for American scientists in the era. Science only was to be their calling.[28]

Because science and art represented similar arenas of competition between East and West for legitimate world leadership in the early Cold War, an analogy from art history can help clarify the geopolitical relevance of basic scientific research in certain key fields, including molecular biology. In a famous 1939 essay that bears comparison with Merton's contemporary work on scientific autonomy, both in argument and in political implications, art critic Clement Greenberg condemned the overtly political, representational art of Hitler's Germany and Stalin's Russia (along with Western commercial art) as "kitsch." Kitsch was a perversion of art, the opposite of truly cutting-edge art that followed the internal logic of each medium's inherent developmental path. In painting Greenberg identified this true path with abstract modernism, where an apolitical elite of artists' artists self-consciously struggled for genuine cultural progress. After the War, as art historian Susan Buck-Morss puts it, this art-for-art's-sake ideology or "'apolitical politicism' became a weapon in the Cold War, when nonrepresentational art came to be equated with democratic societies." Thus abstraction became the official high art of the Free World, a defiant symbol of the artistic freedom absent from the Communist East. Grandiose abstract canvasses displaying simple squares—in limited palettes, preferably mono-chrome—spilled forth from the brushes of the big names in 1950s painting like Albers, Kelly, Marden, Martin, Reinhardt, Rothko, and Stella, winning the highest art accolades and prizes. The square painting trend reached its climax across the West in the early 1960s, writes Buck-Morss, marking an explosion of artistic US imperialism calling itself "internationalism." A similar point has been made by other historians about modern dance and jazz (e.g., Martha Graham and Dizzy Gillespie served as cultural emissaries, both selected by high-level propaganda strategists for federally financed and promoted overseas tours), and arts and letters generally.[29]

In the same way, molecular biology became a key ornament for the West's cultural-political struggle with Communism in the late 1950s and early 1960s. In this it followed the pattern of the Marshall Plan to rebuild Western Europe (which included funding to keep European scientists from looking East, and which in 1951 was already conceived in Washington as part of a "political warfare" campaign in which "truth is our weapon"), and of Eisenhower's Atoms for Peace program to showcase peaceful physics research in the mid-1950s. NATO, too, housed science funding programs designed to build "trained manpower for freedom" in Europe from 1957. Thus, it was crucial ideologically and diplomatically that molecular biology, a basic science with intellectual prestige partly borrowed from physics and little military utility, be well represented in the great universities of the Free World. All the more so since it drew special political significance as an offshoot of genetics—the science most notoriously subject to suppression by Stalin. This was the field that put the gene, whose very existence Communism long denied, in a test tube for all to behold and cast it as life's prime mover. Moreover, genetics and molecular genetics were presented to the world as fields in which the United States was particularly distinguished—perhaps the first where Americans surpassed Europeans. So *Time* magazine suggested in its mid-1958 feature "Heredity—life's hidden mechanism," picturing Beadle on the cover of both domestic and foreign editions (and comparing Beadle's apolitical stance on atomic testing favorably with Pauling's vocal opposition). Molecular biology's status as an exemplary science for science's sake, together with its atomic and anti-Stalinist resonances, help explain the political and cultural favor it enjoyed in the late 1950s and early 1960s. Thus, while biophysics in the first postwar decade signified a life-affirming compensation for or benefit from the atom bomb, by the later 1950s the field's main social significance was as an icon of a fundamental science showcasing the free thought that only the Free World supposedly offered.[30]

To the Congress that funded it, molecular biology and other basic life sciences could simultaneously serve another more explicit purpose, namely demonstrating a commitment to improving the health of Americans by promoting pharmaceutical and other medical advances through private enterprise, rather than through President Truman's national health insurance scheme that the American Medical Association and allies narrowly defeated around 1950. In drawing funding from NIH, molecular biologists—like all life scientists—had to make a case that their research would contribute to understanding the mechanisms of cancer and other dreaded illness so that new cures might be developed. The basic research they did thus consolidated its (literally) preclinical character: it was not just potentially useful, but it explicitly modelled disease and shadowed the clinic, perhaps providing the basis for the next heart drug or cancer radiation therapy—a public-domain resource to supply the pharmaceutical industry and other elements of the American fee-for-service health system. (In contrast public health research, even into crucial topics like the social and environmental contributors to heart disease and cancer, has never received more than a tiny fraction of the funding for clinical and basic biomedical research, particularly before the National Institute of Environmental Health Sciences in NIH began operating in 1967.)[31] Thus basic research in molecular biology offered a double helping—because of the Lysenko constrast—of fundamental science's ideological glamour, and as a bonus-produced knowledge only a step or two removed from practical implementation by the drug and other health industries. No wonder it thrived so well in the early Cold War. When that geopolitical era passed in the early 1970s, molecular biologists would experience many ramifications. Many tried to foreground and own the utility they previously placed in the remote future of their research. Engaging more with practical applications and commerce, they were compelled to become more oriented to the opinions of people other than their scientific peers, and less disinterested. Some would become very political on behalf of their new enterprises. The scientific world Kuhn described as normal would, a mere 20 years later, seem a receding Shangri-La.

Inside Molecular Biology:
Progress of a Normal Science, 1962–1972

Politics aside, with the operon model and Crick's central dogma in hand, and the "genetic code" already largely "cracked," by 1962 the newly recast field of molecular biology had a settled paradigm. The many young researchers who flocked to the field in the 1960s could safely build careers in this well-funded science, filling in the missing pieces of the puzzle of genetic information flow in bacteria (or as biologists said, prokaryotes). Alternatively, they could follow the many leading figures that began in the early and middle 1960s to study higher organisms (or eukaryotes) instead of the microbes that no longer promised big breakthroughs. Turning to creatures well studied by geneticists, Francois Jacob took up the mouse and Seymour Benzer the fruit fly *Drosophila*. With daring and originality, Sydney Brenner began to study the genetics and development of the worm *Caenorhabditis*. Still others turned to animal viruses, expecting the same sort of insights into gene regulation that had flowed from the study of bacterial viruses. Cancer research was especially fundable, and the transformation of animal cells in a culture dish with a cancer-inducing animal virus was a technically and conceptually neat analogue to the attack on a bacterial culture by phage.[32] Thus Rous Sarcoma Virus, which causes cancer in chickens, and the monkey cancer virus SV40 (a contaminant in polio vaccines), became popular objects of experimentation in gene expression.

Much as it had been a decade earlier, the Holy Grail of molecular biology in the late 1960s and early 1970s was the problem of how each genetically identical cell expressed just the appropriate set of genes in the right time and place—or not, in the case of cancer. The operon theory answered for single-celled organisms, but what of higher organisms, whose individual cells somehow turned into groups of distinct tissues and organs? Was there such a thing as a eukaryotic operon? As historian and scientist Michel Morange has insightfully observed, molecular biologists that had never worked on anything but bacteria and phage were surprised to discover that multicellular organisms were complicated and required different laboratory techniques. Breakthroughs were not quickly forthcoming, and the operon model of transcriptional control was not easily extended to many cases of gene regulation in higher organisms. A rash of speculative models of gene regulation emerged from the widespread tendency to overinterpret what little new data became available. For example, the discovery of the reverse transcriptase enzyme from one cancer virus, at first rejected because it clashed with the received view of DNA-to-RNA information flow, was suggested by some as a general mechanism by which cells remodel their genes as they differentiate, turning RNA into new DNA throughout the chromosomes. In any event, by 1970 the most pressing questions in molecular biology, and perhaps all biology, centered on the organization and regulation of genes in eukaryotes. One indication of this importance is focus of the Nobel Prizes in Physiology or Medicine given between 1975 and 1995 (because of time lag, all related to work that either beginning or culminating in the 1970s): six of the seven life science Prizes given for molecular biology in the interval dealt with this topic area.[33]

To biologists engaged in the quest for the imagined eukaryotic operon, and struggling to fit in alternative regulatory mechanisms such as messenger RNA (mRNA) processing and degradation, and protein translational control and degradation, the key lay in finding a particular gene whose expression in a higher organism could be conveniently measured, in order to explore what turned it up and down. Since it proved much more difficult to isolate mRNA from particular organs and tissues of higher organisms than from bacteria, researchers sought specialised cell types that expressed a particular gene at high levels, preferably in tissue culture dishes. Thus cells of the chicken oviduct, which mass-produce the protein ovalbumin (egg white) when stimulated with sex hormones, were one favorite "model system" for eukaryotic gene expression. Another was the globin proteins that make up hemoglobin, churned out by certain blood cells. Once the scientists had identified the mRNA of interest—and even in such specialized animal cells there was lots of other RNA floating about—they could monitor the transcription of the gene. If they isolated large enough quantities of reasonably pure mRNA for the gene, they could copy the RNA into complementary "cDNA" using reverse transcriptase in a test tube. With the improved DNA sequencing methods introduced in the mid-1970s, they could then look at the sequence, including the sequence upstream and downstream from the coding part of the messenger RNA, which presumably controlled the way the protein was produced from it on the ribosomal machinery.[34] And with the new techniques for molecular "cloning," or cutting and pasting foreign DNA into live bacteria (discussed below), which emerged around 1973, they could also use the cDNA to find the original gene in the animal chromosome, clone it, and look at the adjacent DNA further upstream for sequences that determined when the genes were transcribed into mRNA. Scientifically, this was the most exciting prospect that became available hand-in-hand with the technical advances paving the way to commercial genetic engineering. But this is to get slightly ahead of the story.

While the eukaryotic operon continued to evade its seekers throughout the

1960s, curiosity-driven research on bacteria and their viruses continued to reward those who stuck with the purist program, and led in many directions. One was a surprise that grew out from an observation in the 1950s that some bacteriophage could only grow on particular strains of the bacterial species they parasitised. This host strain "restriction" became a lively topic of investigation. The Pasteur group discovered, using radioactively labelled phage, that restricted viruses attacked resistant bacteria normally, but that the phage chromosome was broken up when it entered the host cell. Some enzyme produced by the bacterium evidently served as a defense mechanism against foreign DNA. Clever experiments, particularly by Werner Arber at Geneva, further showed that bacteria with the defensive enzyme also produced another enzyme that modified their own chromosome, so that it would not be attacked by their "restriction enzymes." Bacterial mutants were found that lost the ability to attack virus DNA, but that still conferred protection on their own (and phage) DNA from attack, thus distinguishing the genes for the restriction enzyme and for the modification enzyme that blocked its action. By the early 1970s, several bacterial restriction enzymes had been purified enough to study the DNA-cutting reaction in vitro. As a group at Johns Hopkins discovered, one type of restriction enzyme (Type II), initially found in the bacterium *H. influenzae,* cut unprotected DNA into only a few pieces, strictly at a specific target sequence.[35]

In 1971 another Hopkins group used this newly isolated enzyme to cut pure SV40 chromosomes, passed them through a gel electrophoresis tube to separate the 11 resulting pieces by size, and drew a map—a "physical" rather than genetic map—of the order in which these pieces or restriction fragments stood on the intact chromosome. Again, in a trajectory that may be common in all experimental sciences, the once-obscure object of investigation was recycled as a standard laboratory tool. "Suddenly, everyone wanted to map DNA and use any available restriction enzymes to examine their favorite genome," recalled Richard Roberts, then a restriction enzyme researcher at CSH. "At first, we gladly shared these enzymes with the academic community, but the demand quickly became far too high." Only a few years later, as interest in these basic instruments of genetic engineering continued to skyrocket, Roberts went on to join the Boston-area reagent firm New England Biolabs, which began selling the enzymes in 1975.[36] Many such firms would soon spring up, replacing the old gift economy of molecular biology with a service industry supplying the unfolding boom in genetic engineering.

Another key methodology of genetic engineering emerged in the early 1970s from basic research on the molecular genetics of bacteria and phage. Around

1960 investigations of RNA's role in gene expression, in particular those using virus infection to trace gene expression events, required a technique to identify strings of nucleic acids (RNA or DNA) that came from phage rather than host bacteria. Single stranded isotope-labelled RNA or DNA from phage-bacteria mixtures could be added to pure phage DNA labelled with a second isotope. The intertwined strands of double-stranded DNA were separated by heat, the mixture allowed to mingle under warm conditions, and then cooled. Treated thus, newly formed double-stranded "hybrid" nucleic acids could be identified as viral by the presence of both isotope labels (together with gross physical properties indicating double-strandedness). Hybridization meant that the two strands shared a great deal of complementary sequence, in the same way as the two perfectly complementary strands of a DNA in a chromosome. At first the method of mixing a specific labelled nucleic acid strand with another to see if the two were complementary (i.e., shared many gene sequences) was inconvenient, requiring several days of ultracentrifuge time to produce a density gradient that could physically separate double-stranded from single-stranded nucleic acids. In the early 1960s the ultracentrifuge step was eliminated by conducting the entire process of heating and cooling on a physical support to which DNA was attached, so that the unhybridized nucleic acids could be simply washed off. The best method proved to be a certain filter paper made of nitrocellulose to which DNA stuck permanently when baked.[37]

Despite such improvements, hybridization experiments could only be performed pairwise. Still, throughout the 1960s the technique proved quite useful. For example, it allowed animal virologists to prove that certain cancer-causing viruses had chromosomes made of RNA that were turned into DNA in infected cells (as in the retrovirus hypothesis), and it enabled evolutionary biologists to measure the relatedness of different species via the kinetics of hybridization between their DNA sequences. At the beginning of the 1970s, though, the same refinements in gel electrophoresis methods that improved researchers' ability to separate nucleic acid (and protein) strands according to their length similarly made hybridization much easier and more useful. Gel electrophoresis capitalizes on the fact that DNA and RNA have a negative charge, and move toward the positive electrode at the far end of a solution or gel when voltage is applied. The smaller pieces pass quicker through the narrow pores of a gel matrix than the larger pieces, so that a heterogeneous collection sorts itself by size. Now a chromosome or other large piece of DNA could be cut into a number of tractable-sized pieces with restriction enzymes, and the pieces separated into bands reflecting their lengths on an electrophoresis gel. There they could be visualized, either immediately with stains if the sample was plentiful, or more slowly via radioactive label and X-ray film. And the bands could be transferred directly to a nitrocellulose filter sheet simply by pressing it to the gel. This DNA band-bearing sheet or "blot" could then be rinsed with a solution containing another, radioactively-labelled "probe" sequence of DNA or RNA, which would hybridize with and thus adhere to any band containing DNA with the complementary sequence. Called a Southern blot after its inventor, the procedure can answer not just *whether* one DNA string contains sequences similar to those on another, but *where* on a mapped chromosome or other long DNA the sequences matching the probe are situated.[38]

The Advent of Recombinant DNA

The idea that molecular genetics, in its procedures for isolating particular genes and moving them into foreign organisms, offered practical promise and profound danger predates even the restriction enzymes in the genetic engineer's toolbox. In late 1969 the Harvard Medical School team of molecular geneticist Jonathan Beckwith published a high-profile paper in *Nature* describing how, using a combination of classical phage genetics techniques and both physical and enzyme manipulation of DNA, they isolated the *E. coli lac* operon in "pure" form (the repressor binding site, promoter sequence from which transcription begins, and entire first gene encoding the B-gal enzyme). While the immediate use of this DNA was mainly to study regulation of the *lac* operon in the test tube (for instance, binding studies with the purified repressor protein), the extensive news coverage given the report stressed both the perils and the hopes around human genetic engineering. For instance, in the *Washington Post*'s front page story "Research Isolates Gene for First Time," Beckwith and junior Harvard colleague James Shapiro said that "they fear the day when other scientists, using the techniques they have developed . . . may seek to engineer human beings according to some government's specifications," even though their discovery could also be used beneficially to "eradicate genetic diseases and deformities." The scientists related their fears to the uses of biological and chemical technology in the Vietnam War and the military's prior use of physics to develop nuclear weapons. They were deeply concerned lest molecular geneticists soon find themselves "a group of very regretful Oppenheimers."[39] A few years later this curious combination of self-promotion and self-denunciation would be reprised, in somewhat less exaggerated form, by the mainstream molecular geneticists who called for a moratorium on their own work. And as we will see, the perspectives of Beckwith and his left-leaning colleagues would also resurface when the recombinant DNA debate erupted beyond the biomedical community in 1976.

What is now regarded as the advent of genetic engineering came in a set of 1972 experiments by the lab group of Stanford's Paul Berg. While aiming at a basic biological question, how bacterial genes might be expressed in animal cells and vice versa, the work was most important for its methodological innovation. In brief Berg showed how the techniques already in play among various molecular geneticists could be deployed, as an ensemble, to mix and match gene segments in vitro and then move them between living species. Using enzymes provided as gifts by five different laboratories, Berg's group opened the circular chromosome of virus SV40 by cutting it once with a restriction enzyme, did the same to a bacterial "plasmid" (or autonomous mini-chromosome) containing the *E. coli* galactose operon, digested away some of one strand on the chromosome ends with another enzyme, added a sticky "tail" (consisting of a string of adenine bases, or their complementary thymidines) with yet another, and joined the pieces together with ligase into circular loops containing both the SV40 and the bacterial DNA (see fig. 1.1). The result was a "biological reagent": that is, a vehicle for growing genes of interest, with their on-off switches, in mammals or in bacteria or shuttling back and forth between the two. Berg's 1972 paper is rightly recognized as a tour de force and a "work of art" by scientists, in that it clearly points the way to genetic engineering as a general enterprise through cutting and pasting with the newly understood Type II restriction and other enzymes.[40] It also raised alarm. One plausible reason (though later disavowed and evaded by Berg and most mainstream biologists) was that the monkey virus DNA could be used to move foreign genes into animals—and higher apes in particular.

By 1970, times were changing, both geopolitically and in the microcosm of science. On university campuses, students occupied buildings demanding an end to military-funded research at their institutions. Scientists were urged to take political responsibility and pursue socially "relevant" research rather than, through apolitical science for its own sake, passively support the national security state (or "Establishment war machine," as campus radicals put it). After two decades of expansion, the de facto American science policy of limitless federal support for basic research at universities faltered. The Mansfield Amendment of 1969 responded both to Congressional cost concerns and the unpopularity of the Vietnam conflict by curtailing military agency funding of basic research: having supported as much as one-third of university research at the beginning

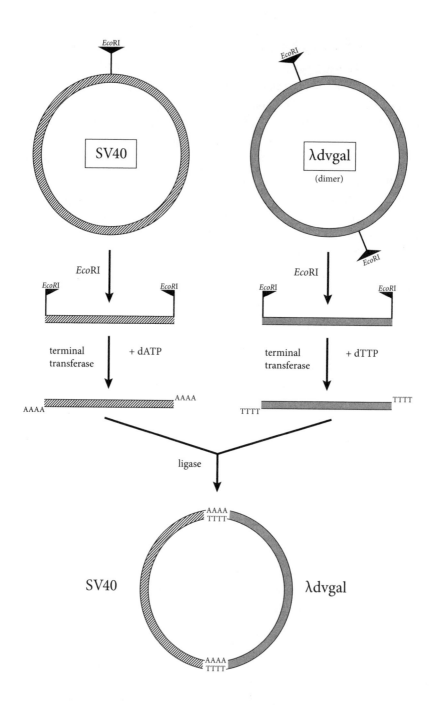

of the 1960s, the Defense Department accounted for 15% of university research dollars in 1970 and a paltry 8% in 1975. This cut in military-funded research at universities reflected a major decline in overall defense spending, dropping from 10% of the US domestic product in 1968 to under 5% in 1979, signalling a lull in two decades' of incessant Cold War and the opening of an economically rationalist coexistence with the East, détente.[41]

Still more relevant to biology, in 1970 the NIH budget dropped 10% in real terms. Early that year a rumour that NIH might de-fund all work on *E. coli* and its phages stirred panic among molecular biologists. The 1971 National Cancer Act (President Nixon's "war on cancer") had the effect—even though it was couched in terms of new funding—of diverting NIH from curiosity-driven work on gene function generally toward research more immediately related to cancer. 1971 brought the NIH budget, in real terms, just back to where it was in 1969. After a 20% jump in 1972 from cancer dollars it barely kept up with the high inflation rates of the next five years. Success rates for grant applications to the NIH wandered downward from their peak near 60% in the early 1960s, to about 40% in the early 1970s (with a very precipitous drop in 1973), to about 30% by 1980, and then lower still. As the biggest funder of life science and the one that in the 1950s had, in the words of Geiger, "tipped the balance of the university research economy from a programmatic to a disinterested orientation," NIH now tipped the balance the other way. Paul Berg had anticipated the changes when he switched his research from bacteriophage to cancer viruses in 1967–68. With NIH funding at a plateau, and NSF not even holding its own, America's ever-expanding pool of life scientists found itself increasingly starved in the mid-1970s.[42] Negative economic growth did nothing to encourage faith in a speedy return to the good days of a scientifically magnanimous national security state. The ideology of basic research that had governed molecular biology began to dissolve as scientists looked harder for immediate medical and industrial justification, and support, for their research.

The impetus to work on problems more relevant to human health, however,

---

Fig. 1.1. Construction of the 1972 product of Paul Berg's lab that prompted the recombinant DNA moratorium. A complete genome of SV40 (a cancer-causing monkey virus) was linked to a complete plasmid capable of replication in bacterial cells, by cutting each with a restriction enzyme and rejoining via A-T tailing. (Not all enzymes and substrates depicted; see source and main text for further details.) Abbreviations: A, adenine; dATP, deoxyadenosine triphosphate; dTTP, deoxythymidine triphosphate; T, thymine. *PNAS* 1972;69:2904–9.

raised its own social hazards. As we saw in the case of Harvard's Beckwith, both the concept of genetic engineering and its dangers cropped up even before the ready availability of Type II restriction enzymes. Among molecular biologists certain DNA-recombining experiments raised special alarm because they involved genes from animal viruses. The work with monkey cancer virus SV40 being planned by Berg and his colleague Janet Merz was a particularly frightening prospect. Mixing and matching genes with *E. coli,* which already infects humans, could accidentally give the bacterium more dangerous infectious traits acquired from SV40, or worse yet, create a new super-virus causing cancer in humans. Because of these worries, Berg's group did not actually infect any living bacteria or monkey cells with his SV40-plasmid chimeric DNA. Berg and a few others raised the alarm sufficiently to put such experiments on hold, spurring broad discussions among molecular biologists about the dangers of various recombinant DNA experiments. These discussions led to a formal study of the safety problem under auspices of the National Academy of Sciences (NAS), chaired by Berg. On behalf of this NAS committee, in mid-1974, Berg published a prominent call for global moratorium until the NIH and other government bodies could set regulations on most recombinant DNA work.[43]

Berg and his fellow committee members consulted widely within molecular biology and virology. Exchanges between Berg and other scientists help illuminate why this scientific community supported the moratorium and risk assessment process. Berg persuaded the others by invoking prominent scandals in which the public had recently lost trust in scientists: failures of the academic medical research establishment to respect the best interests of research subjects (as in Tuskegee and Willowbrook); the untrustworthiness of the pharmaceutical industry (as illustrated by thalidomide); and the complicity of many atomic scientists in disguising the hazards of nuclear test fallout.[44] As for initial views on what recombinant experiments presented the greatest risks of public health and publicity disaster, the biological community flagged as particularly dangerous not just experiments introducing genes from cancer viruses to bacteria, like Berg's, but also experiments introducing any DNA that might contain latent animal virus genes into bacteria. Because latent viruses were thought to be mingled among the chromosomes of humans and other animals, this concern effectively ruled out gene splicing experiments using animal DNA, unless the genes on the DNA in question were known to be safe. As yet, very few genes from higher organisms had been characterized. According to mainstream opinion, then, biologists had to be particularly cautious with a type of experiment that presented itself as obvious and technically sweet: cloning of genes from humans and other higher animals into bacteria by the "shotgun" approach (so called because it involved randomly introducing animal DNA fragments into living bacteria, and afterwards selecting those carrying genes of interest).

It was during this self-imposed moratorium that the most famous early demonstration of genetic engineering, by University of California, San Francisco molecular geneticist Herbert Boyer and Stanford microbiologist Stanley Cohen, neatly sidestepped both safety concerns and technical obstacles. In 1973–74 Cohen and Boyer conducted a series of experiments that culminated in moving a frog gene into a bacterium using the *Eco*RI restriction enzyme studied by Boyer's lab—the same restriction enzyme Boyer had given Berg for his 1972 experiment. Many other key enzymes and materials were the generous gifts of molecular biologists elsewhere, as was still standard practice in the day. Using *Eco*RI the team inserted the gene for frog ribosomal RNA into a plasmid provided by Cohen, and then showed that the recipient bacteria could read the introduced frog gene into RNA—effectively rehearsing the basic operation needed to produce animal proteins in bacteria (as news coverage pointed out), only without the protein. Both Boyer and Cohen were signatories on the Berg moratorium letter. This project was arguably safe because the gene for ribosomal RNA, one of the few that had been isolated, had a very well-known, innocuous function in the generic protein synthesis machinery. Furthermore there was no reason to think frog ribosomal RNA would give bacteria a survival advantage either in people or the environment, even if the foreign RNA ended up functioning as part of the bacterial protein synthesis machinery (which seemed unlikely). So, presumably, no danger.[45]

The Cohen-Boyer work differed from Berg's pioneering paper mainly in using the naturally sticky ends left by a restriction enzyme on cut DNA (instead of added complementary A-T tails) to insert the foreign gene. Based on this distinction, in late 1974 an especially enterprising Stanford intellectual property lawyer filed and eventually obtained patents on the basic process of inserting and expressing foreign genes in bacteria using restriction enzymes and a plasmid "vector" (carrier). Boyer and Cohen were thus honored as the inventors of genetic engineering technique in patent law, and hence in the business world. But they were not quite so honoured by biologists, many of whom saw the actions of Cohen, Boyer, and Stanford as less than fully admirable (in that, as one put it, they were patenting "procedures derived from work done during the moratorium by people calling for it"). To some, deviations from the ideology of basic research such as patenting still seemed inherently dishonourable. At the outset Cohen responded to the situation by relinquishing his share in the Stanford patent royalties to general research funds, so as to show he was not motivated by "personal gain or aggrandizement." Boyer, who in 1974 was already thinking about engineering bacteria to make pharmaceutical hormones, did not do likewise (although he later donated royalties to UCSF).[46] In 1980, Berg alone received a Nobel Prize for genetic engineering method—shared with the inventors of methods for determining DNA sequences, Walter Gilbert of Harvard and Fred Sanger of the Cambridge LMB.

In February 1975 Berg's NAS committee hosted a large international conference in Asilomar, California, for biologists to establish safety rules for recombinant DNA work. Sixteen journalists were invited to attend, and of course there were also officials from the NIH, whose guidelines the meeting was designed to recommend (although there was widespread expectation that the rules would extend internationally). Among more radical critics only Beckwith was invited, but he could not attend. The Genetic Engineering Group of Boston's left-wing Science for the People sent instead a lengthy, carefully crafted open letter to all scientist attendees, raising concerns about the broader social implications and also calling for the participation of laboratory worker and community representatives in NIH panels assessing genetic engineering experiments. This letter was entirely ignored. Molecular biologists committed to genetic engineering dominated the meeting, and they limited discussion to the safety assessment of different classes of gene transfer procedures, mainly in lower organisms. That was the type of work they were especially eager to proceed with. The goal in the minds of many of these scientists was "to give the public some assurance" that they were exercising due concern for safety (as one put it), and then to get on with what they considered the safer experiments.[47]

The 1975 Asilomar meeting resulted in classification of gene transfer work into categories, which with minor changes became NIH rules; effective in June 1976, their official issue ended the moratorium. Especially stringent, or so it soon would seem to scientists, was the restriction of shotgun experiments with DNA from mammals and birds to P3 labs—facilities offering the second highest degree of physical containment—and specially enfeebled bacterial strains. (Shotgun cloning of adult human DNA was then restricted to P4, germ-warfare style labs, in practice virtually banning it.) P3-standard labs were uncommon and more expensive than typical hospital or university microbiology labs at the time, largely because they required negative air pressure and special ventilation to keep microbes from escaping the room alive. From the Asilomar meeting until the 1976 end of the moratorium under the NIH guidelines, however, it seemed to mainstream molecular biologists that such constraints imposed by their cautious self-reflection could be accommodated, and were worth the benefits of avoiding both stricter regulation and a new plague of their own creation. Certainly, the assumption that this their exercise in self-regulation entitled them to clone without further interference surfaced later in 1976 when, faced with new opposition mainly from outside the scientific community, some biologists expressed regret that they had ever brought the potential risks of gene splicing to public attention.[48]

Boyer was just a little ahead of his time among molecular biologists in appreciating the way the subject matter of the field had turned into a kit for making bacteria manufacture drugs. In 1974 he was already thinking about engineering recombinant bacteria to produce hormone pharmaceuticals (including human insulin and angiotensin-2) and had even tentatively contacted a big drug company. Thus he was prepared to listen when 28-year-old businessman Robert Swanson approached him in January 1976 with ideas about the commercial applications of genetic engineering. Swanson had apprenticed in venture capital, a then-obscure type of finance flourishing in California around the electronics and computer industries. Venture capitalists invest a few hundred thousand dollars to get a company started in exchange for a large part of the equity, and then help manage the firm's growth for a few years until it matured enough to sell their part to a large corporation or on the stock market. Failure is frequent, and high rates of return are demanded to compensate for the high risk.[49]

By the end of the day they met, Swanson invited Boyer to co-found a company to commercialize recombinant DNA technology, and they quickly drew up a short list of protein hormones with commercial potential. In April 1976 the company incorporated and had already embarked on a business plan to clone its first two hormones. Human insulin would be their first commercial target. From the start Genentech had competition nipping their heels—molecular geneticists going after the same protein drugs. It took no supernatural insight to sense the opportunities, intellectual and commercial, presented by genetic engineering techniques. Indeed, as incoming FDA Commissioner and Stanford biologist Donald Kennedy quipped in March 1977 at the National Academy of Sciences forum on recombinant DNA safety (see chap. 2), "I don't know . . . much about recombinant DNA, . . . but I surely can tell when a party is on its way."[50]

CHAPTER TWO

The Insulin Trophy

The first pharmaceutical that molecular biologists made was human insulin. It was always going to be insulin. From the perspective of life scientists in the middle 1970s, it is hard to imagine any other drug coming sooner into the medical marketplace through genetic engineering. This statement is not intended as an endorsement of certain dominant stories about biotechnology, rightly criticized for their technological determinism: the view that the state of technique in molecular genetics of the mid-1970s, together with the logic of markets, have mechanically and inevitably given us the present list of biotechnology drugs. The determinism that I am invoking is as much cultural as technological, and decidedly uneconomic. In the mid-1970s, there was an adequate supply of the slaughterhouse-derived pig insulin that had served medicine for half a century; projections of a shortage were remote and disputed. For the few diabetes patients that developed allergies to animal insulin, a humanized insulin made by converting amino acids chemically was far along in development.[1] So medicine did not then require recombinant human insulin.

Rather, the scientists picked insulin. For the multiple entrepreneurial biologists who simultaneously decided to clone this protein hormone, it represented the project best suited for exploiting a newly emerging congruence of business and scientific interest, and for cultivating and expanding that zone of overlap between what could bring scientific credit and commercial reward. Cloning the human insulin gene into bacteria was expected to gain fundamental new knowledge in molecular genetics, since the effort to express the gene might further elucidate mechanisms of transcription and translation in bacteria, and also might afford a look at the mechanisms regulating genes in higher organisms (e.g., sequences in mRNAs controlling translation, and perhaps the nearby sequences regulating transcription—the "eukaryotic operon"; chap. 1). But for these intellectual goals, the gene for any well-characterized plant or animal protein would do. Basically, the scientists picked insulin because insulin was the "most famous protein" (as one biologist put it).[2] That is, insulin's rich patina of glamour explains its mutual attraction for molecular biologists seeking intellectual distinction, and for all the many stripes of entrepreneur seeking practical mastery of genetic engineering.

To be sure, biologists knew as much about insulin in the middle 1970s as any protein—and this same history of increasing knowledge was a history of accumulating glory. When insulin first became available commercially in 1923, through a collaboration between the Lilly drug firm and University of Toronto researchers Banting, Macleod, Collip, and Best, it was immediately hailed as a miracle drug for saving lives among the roughly one-half percent of the population who were diabetic. This success inspired emulation and spurred the interwar hormone gold rush.[3] In 1926 the "father of American pharmacology," J. J. Abel, made insulin's crystallization—for chemists at the time, crystallization represented ultimate purification—the capstone of his career. Insulin was only the second protein ever analysed by X-ray crystallography, by British biophysics pioneer Dorothy Crowfoot in the mid-1930s. In 1955, insulin was the first protein whose amino acid sequence was completely determined, crowning a decade's intense work by biochemist Frederick Sanger at Cambridge.[4] Insulin's trail was paved with Nobel Prizes: Banting and Macleod's in 1923, Sanger's in 1958, Crowfoot's in 1964 (largely for other work; she had not yet finished studying insulin). Essentially, insulin symbolized the fruitful conjunction of advanced biological research with medicine, through scientific drug development. The first recombinant drug had to be insulin because the molecular biologists made it their trophy.

Cloning Rush Hour

The very first gene for a protein from a higher animal cloned into bacteria by genetic engineering was not insulin but globin, the protein part of hemoglobin. (Cohen and Boyer's frog ribosomal gene was a gene whose product was not a protein but the RNA itself; chap. 1.) Molecular biology's intellectual and technical logic made globin one of a very few candidates for the honor. Globin was the largest constituent in red blood cells and a favorite object of study by protein biochemists for nearly a century, so it was well understood. Its choice as the first protein-coding gene from higher organisms to be cloned into bacteria stemmed from a technical constraint of the mid-1970s. Molecular biologists then could not simply reach into the chromosomes of higher organisms (i.e., eukaryotes) and pull out the DNA for a desired gene, as they were learning to do with the simpler bacteria. The chromosomes were too many, too big, and mostly uncharted by traditional breeding-based genetics. To obtain the DNA encoding a particular eukaryotic protein, the only plausible route at the time was indirect: first to obtain the messenger RNA (mRNA, or message) from cells actively producing the protein, and then to produce a DNA copy (cDNA) of that mRNA by treating the RNA with the recently discovered enzyme reverse transcriptase (from certain cancer-producing viruses with chromosomes of RNA instead of DNA, the "retroviruses").[5] That cDNA copy of the gene could be spliced into a plasmid with restriction enzymes, and then introduced into living bacteria. Because of the low efficiency of each step, biologists needed an abundant mRNA, preferably one not mixed with mRNAs coding for other proteins. This is what made globin ideal: red blood cells produce hardly any other protein, and blood is easy to get. Messenger RNA extracted from red cells would be nearly pure globin mRNA.

Such, at any rate, was the thinking of Harvard PhD student Argiris Efstratiadis and the young Harvard professor Tom Maniatis. They started work during 1975, under James Watson's wing at Cold Spring Harbor Labs on Long Island, following the informal Asilomar guidelines before they were made official.[6] Efstratiadis's part was isolating the globin message and making it into cDNA. Working with rabbits, it was straightforward enough to bleed the animals, spin the red cells out from the serum, and extract RNA simply by breaking up the red cells in the presence of phenol (a corrosive, carcinogenic benzene derivative that disrupts enzymes and other proteins)—all in essentially the standard way biologists extracted DNA and RNA from bacteria. After running the samples through procedures to clear them of protein and DNA, and to separate mRNA from ribosomal RNAs, globin message could be seen as a single, prominent band on a size-separating electrophoresis gel made with formamide (another corrosive chemical, used to keep the RNA from folding). All this had been done before.[7]

It was a simple further step to cut out the band from the gel and extract the globin message from it. Treating the globin mRNA with reverse transcriptase enzyme, either before or after running the gel, made globin cDNA: a quasi-artificial gene, a piece of DNA mirroring the sequence of the natural globin gene as transcribed by the blood cell into message. With terminal transferase, another of the seven enzymes donated by other scientists for this experiment, Efstratiadis and Maniatis used A-T tailing to insert the cDNA into to the Boyer lab's pMB9 plasmid, in the same way as Berg. The recombinant pMB9 plasmid containing the rabbit globin sequences was then successfully moved into live *E. coli*. Efstratiadis and Maniatis completed their rabbit globin cloning at about the same time that NIH announced its draft guidelines in February 1976, superceding Asilomar. This work made a big splash scientifically: it showed not only that one could clone a specific gene from mRNA but also, by reading the sequence of the cDNA clones, that these were faithful, one-to-one copies of the RNA.[8]

Efstratiadis and Maniatis had shown that one could clone genes from isolated mRNA, but they had not expressed their cloned globin gene as a protein; the rabbit DNA sequence just sat there in the bacterium. In the first part of 1976 Efstratiadis was recruited as a postdoctoral fellow by Harvard biology professor Wally Gilbert—inventor of the new Maxam-Gilbert DNA sequencing method— to achieve expression. But despite Efstratiadis's skill with rabbit globin, Gilbert had another target in sight. Gilbert planned to clone insulin, and to produce the protein hormone in bacteria. Expression of cloned globin in bacteria was left to other lab groups, such as that of Harvard colleague Mark Ptashne. As Stephen Hall has described it in his excellent book *Invisible Frontiers,* Gilbert's decision to produce insulin in bacteria typified the attitude he cultivated in his laboratory group, where the effort was always to stun the world with a breakthrough, rather than rest content with pedestrian, incremental progress.[9]

Efstratiadis was to isolate the insulin mRNA and to make cDNA from it, following much the same procedure as with globin. He would then provide the insulin cDNA to another member of Gilbert's Harvard team, PhD student Forrest Fuller, for cloning into a vector specially prepared to enable the cDNA's transcription into mRNA and translation into protein. Expressing a cloned protein— as opposed to merely inserting it into a bacterial plasmid—represented a major additional challenge beyond the globin work. First, the cDNA would have to be trimmed to remove sequences other than those encoding the desired protein. These would include the noncoding parts of the mRNA that come after the natural promoter from which transcription began, as well as the first part of the insulin protein as initially translated (which is processed off in normal folding and hormone secretion). This trimmed cDNA sequence then had to be positioned a certain distance downstream from a bacterial promoter sequence, from which the bacterium would begin transcribing mRNA (chap. 1). Between the bacterial promoter and the coding cDNA, there had to be a bacterial ribosome binding site at just the right position, and also a three-base codon specifying methionine, the amino acid with which bacterial ribosomes begin translating every messenger RNA into protein (the initiation codon).

And this list of difficulties for achieving expression glosses over many others (such as extra problems because insulin in active form is actually two linked protein chains). The goal could have been approached by multiple stages; for exam-ple first cloning insulin-coding cDNA into the middle of a bacterial gene, so that it would make a fusion protein consisting partly of the bacterial protein and the remainder insulin; and then later reducing the bacterial portion by steps only to the initiation codon. But Gilbert's insulin cloning strategy was designed to grab the brass ring in one master stroke. Fuller's vectors embodied this bold strategy. Starting with pMB9 (and later the Boyer group's improved plasmid pBR322), Fuller inserted a piece of DNA from the Jacob-Monod *lac* operon, molecular biol-ogy's paradigmatic expression system (chap. 1), carrying a strong "up" mutant promoter and ending early in the coding region of the next gene downstream, *lacZ* (encoding the enzyme beta-galactosidase). The *lac* promoter was of course intended to drive transcription. If Gilbert wanted only to express a fusion pro-tein built from insulin and beta-galactosidase, Fuller's initial construct would suffice: onto that sequence for the *lacZ* stub they could simply splice an insulin cDNA, shortened so that its coding sequences continued those of *lacZ* (e.g., by restriction cleavage of a site in the leader sequences in the prehormone, preced-ing the portion coding the mature hormone).

Instead, Fuller did something more ambitious: he constructed a large fam-ily of what he called pOP expression vectors, the *lacZ* sequence of all shortened by judicious digestion with exonucleases, enzymes that chew away nucleic acids from the ends. One of the pOPs ended right at *lacZ*'s initiation codon. Then while Fuller was waiting for enough insulin cDNA from Efstratiadis, he prac-ticed cloning bacterial genes (and also globin DNA, working at Cold Spring Har-bor) into his pOP vectors by blunt-end ligation. This splicing procedure, which did not rely on the stickyness of A-T tails or the staggered cuts left by restriction enzymes to help loose ends find each other, was finicky and inefficient but it was necessary for direct expression: a sticky tail sequence (or added restriction site) between the initiation codon and the insulin sequence would create an abnor-mal insulin protein, or prevent translation altogether. But if insulin cDNA was itself shortened by careful exonuclease digestion to its coding sequence, made into a blunt end, then joined to the pOP vector's blunt end so that an initiation codon immediately preceded the insulin sequence, then Fuller's first bacterial clones carrying insulin cDNA might actually produce insulin protein (fig. 2.1). (Fuller's bacterial clones would likely be making the insulin protein as first pro-duced in the pancreas, called pre-proinsulin, containing extra amino acids that are removed after the protein folds to its correct shape.)[10] The Gilbert group's approach as reconstructed here, the plan to splice a cDNA that began with insulin's coding sequence right onto an initiation codon, can be called the exonuclease strategy for achieving expression.

Obtaining sufficient insulin mRNA for Gilbert's ambitious plan, in high enough quality to clone, presented Efstratiadis with much bigger problems than isolating globin mRNA. First there was the conjoined problem of abundance and purity. Insulin is only produced in quantity by the pancreas, but pancreas tissue produces a far more complex set of proteins than blood cells. At most, a few percent of mRNA from pancreas tissue specifies insulin. There also was also the distinct problem of RNA quality. RNA is much more sensitive to decay than either its sister substance DNA, or proteins like insulin. Even when handling mRNA from blood cells, researchers had to wear gloves and masks and work in formamide and with specially acid-cleaned glass because RNA—attacked both by commonplace chemical reactions as well as a ubiquitous, nearly indestructible RNase enzyme found on skin—has a habit of evaporating in one's grasp. But with pancreatic tissue, the problem was multiplied because the organ consists mainly of tissue loaded with RNA-attacking digestive enzymes, among which are scattered small islands of insulin-producing beta cells.

At first Efstratiadis cut pancreases from rats, grinding the tiny organs in formamide and other chemicals to preserve the RNA. From about June 1976, the Gilbert group made progress on the quantity front thanks to a collaborator Gilbert found at Harvard medical school, William Chick. Chick studied a rat pancreas cancer that grew big on the animals; he supplied the Gilbert team with 20-gram pieces of frozen tumor, each equivalent to more than 40 rat pancreases. However, although each insulinoma tumor produced lots of insulin, the tumor tissue proved no richer proportionally in insulin messenger RNA. And it was certainly no easier to extract that RNA; the tumor tissue was so tough that it had to be ground up with sand and a mortar and pestle instead of the time-honored blender method, all the while worrying about destructive RNAses. For half of 1976 and most of 1977, Efstratiadis repeated the tumor mRNA extraction procedure, finishing with a formamide gel, from which he cut the region presumably containing the proinsulin mRNA (400–500 bases in size) and saved it up for later conversion to cDNA. Meanwhile Forrest Fuller practised his own technique, but his blunt-end cloning was not working well enough to risk months' worth of Estrefiadis's material—irretrievably wasted on a bad cloning day. Gilbert grew frustrated with Fuller's slow progress, suspecting that the inconsistent results of his cloning practice runs stemmed from sloppy, inconsistent technique.[11]

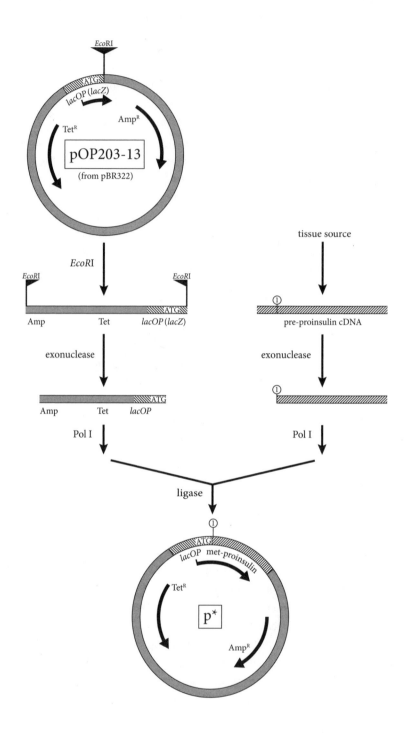

The Crowded Field of Cloners

Gilbert was not the only molecular biologist to set his sights on cloning insulin in early 1976. In October 1975 a 32-year-old biochemist named Axel Ullrich came from Germany, where he had recently finished his PhD, to San Francisco for a postdoctoral fellowship in Boyer's UCSF lab. Another talented postdoc arrived simultaneously: John Shine, fresh from a groundbreaking PhD in Australia, where he determined the sequence of about six nucleotide bases that bacterial messenger RNAs need to lock onto a ribosome. This sequence, usually occurring about eight nucleotides before the initiation codon, allows translation into protein to commence. This was a fundamental discovery in molecular genetics, but also a crucial technical insight for genetic engineering, since any foreign protein expressed in bacteria would need what is still known as a Shine-Dalgarno sequence. Like Peter Seeburg, Ullrich's old university friend already at UCSF, Ullrich and Shine had been captivated by Boyer's work with Cohen in cloning the frog ribosome gene. Boyer encouraged Ullrich to come, but had no money or space to offer. So when Ullrich obtained German postdoctoral funding he was assigned space in the lab of Boyer's then-close collaborator Howard Goodman—effectively forming a free-floating team with Shine and Seeburg.[12]

As a PhD student Ullrich had already begun exploring whether genes from higher organisms could be expressed by bacteria, and vice versa; for example, the eukaryotic equivalent of the Shine-Dalgarno sequence was not known, and it was suspected but not entirely certain that like bacterial genes the coding portion of eukaryotic genes always began with a methinone. Ullrich had even developed a special interest in insulin and glucagon, another pancreas hormone, extracting mRNA from chicken pancreas to see if, and how, a bacterial extract

Fig. 2.1. Reconstruction of the insulin cloning strategy pursued during 1976–1977 by Forrest Fuller and colleagues in Walter Gilbert's Harvard lab. Double stranded cDNA derived from insulin-producing tumor tissue (isolated by electrophoresis either after reverse transcriptase treatment or before as an mRNA) is shortened by limited exonuclease digestion to the first desired codon, and joined by blunt-end ligation to an expression vector containing the *lac* operator-promoter (*lacOP*) and the *lacZ* gene similarly shortened to its translation initiation codon (ATG). Screening would be based on identifying transformants expressing insulin protein, e.g., using antibodies. The strategy was not successful. In this example, the first codon of proinsulin follows the initiation codon, but the first codon of mature insulin chain B could take its place. (Not all enzymes and substrates depicted; see *Gene* 1982;19:43–54 and main text for further details.) Abbreviations: Amp, ampicillin; Pol 1, bacterial DNA polymerase 1; Tet, tetracycline.

might translate it into protein in vitro. On arrival in Goodman's lab he had no fixed idea of what project he should pursue, but he wanted it to be important. Goodman proposed what Ullrich only remembers as "some esoteric yeast project." Ullrich counterproposed that he might try making insulin cDNA from pancreas mRNA—just as Maniatis and Efstratiadis had been doing with globin—and cloning it into bacteria by the Cohen-Boyer method. But Goodman thought the idea unlikely to succeed and so, prudently enough, suggested Ullrich try both projects to maximize his chances of having something to show for his time at UCSF. Soon Ullrich was working with diabetes researcher Jerry Grodsky, a UCSF biochemist who studied insulin secretion and had developed assays to measure the presence of insulin protein (useful for measuring levels of extracted insulin mRNA in the planned experiments).[13]

Little did he know that in trying to clone insulin from cDNA, Ullrich had entered into a competition not only with Gilbert's Harvard group, but also with the chair of his own department. Some twenty years later Ullrich recounted the day he found this out in early 1976, at a visiting researcher's seminar at UCSF on translation of insulin mRNA in a cell-free system:

> After the talk, Jerry Grodsky introduced me and said, "This is Ullrich: he wants to clone insulin." And there was Bill Rutter introducing [postdoc] John Chirgwin and saying the same thing. Afterwards, we met in the elevator and Bill said, "We have to talk." And then Howard [Goodman] and I were essentially ordered into Bill's office and told, "If anybody in this department clones insulin, it's either in my lab or in collaboration with me."[14]

William Rutter ran a large lab group of 20 to 30 junior scientists working on three major projects, one of which was gene expression in the pancreas. Although he had not much background in molecular genetics, he had begun pursuing the cloning of insulin as early as 1975, on the heels of the Cohen-Boyer experiments, by bringing in postdocs with relevant expertise. For a while Rutter tried to tackle both the quantity and quality issues that had been plaguing Gilbert's group by seeking a good source of fish pancreas RNA. Fish, it turns out, have insulin-secreting pancreas tissues separated from the parts of the pancreas dedicated to digestive enzymes. Another attempt to circumvent the problem involved a partnership with a young Texas scientist named Peter Lomedico, who had independently begun his own effort to clone insulin using pancreases collected from cow fetuses. (There is a stage in cow development where the digestive tissues of the pancreas are relatively undeveloped, but insulin is already being made.) This plan fell apart when Lomedico, who had won a new grant for working on the cloning of insulin, cancelled his arrangement to join Rutter at UCSF and went to Harvard as a postdoctoral fellow instead.[15]

Competition with Gilbert's group—underscored by Lomedico's defection—turned up the pace of work at UCSF in the second half of 1976. Whatever plans the California group may have had to approach the cloning of insulin in a way that would also address the problem of expression disappeared. From all of UCSF's subsequent efforts, it appears that their top priority was simply to clone the hormone gene first; making the insulin gene do anything once it was in bacteria would have to wait. The situation *within* UCSF remained highly competitive too. Ullrich was assigned by Rutter to making cDNA from dog pancreas under Goodman's nominal supervision, while Chirgwin worked on rat pancreas directly under Rutter. Meanwhile another member of Rutter's lab, Ray Pictet, worked with Chirgwin on overcoming the RNA quality hurdle with a chemical solution. Through the end of 1976, Pictet and Chirgwin worked on concocting a new recipe for RNA extraction that would slow the RNAse enzymes enough to give insulin mRNA a chance.[16]

Spectator Interference

Throughout 1976 the Harvard team was working in the isolation of Long Island, where Cold Spring Harbor offered labs meeting P3 specifications to accommodate cloning of mammalian genes under the new NIH's new guidelines. This was because Harvard lacked P3 labs—and when the University tried to build one, Cambridge, Massachusetts, banned recombinant experiments. In June that year an internal controversy over the University's plan to build a P3 facility erupted into the public arena. Faculty members including biologist Ruth Hubbard and others associated with the Science for the People group (see chap. 1) opposed the idea of cloning on campus both for ideological and safety reasons. Expecting the university to charge ahead regardless of internal dissent, Hubbard invited a Cambridge City Council member and journalists to a Harvard research policy meeting on the proposed lab. The resulting story in the *Boston Phoenix* weekly, "Biohazards at Harvard," incensed locals—especially Cambridge Mayor Alfred Vellucci—with the University's seemingly high-handed effort to build a deadly germ lab under their noses. At a legendary June 23 special meeting of the City Council, the mayor raked Ptashne and other Harvard scientists over the coals before some 60 television cameras, along with NIH scientist Maxine Singer (a key organizer of the recombinant DNA moratorium and architect of the new guidelines), and other dignitaries supporting Harvard's plans. Why had Harvard not bothered consulting the local community, whose health and safety were at stake? Why should the community trust the assurances of scientists that the research was safe, when the same scientists had vested interests in its progress? Why should the NIH be trusted to promulgate regulations protecting the public from the hazards of such research—any more than the discredited Atomic Energy Commission could be trusted with protecting the public from the radiation hazards its own research created? These were hard questions for which the mainstream biologists, accustomed to conducting their publicly funded basic research without questions from the general public, were unprepared to answer.[17]

Academic critics on the Left, like MIT biologist Jonathan King, were as concerned about large issues of democratic process and ethics (like human genetic engineering and genetic determinism) as they were about microbiological safety. They strategically relied heavily on safety concerns to win public attention, and safety was certainly Vellucci's main theme. Still, the outspoken few were not alone on the safety issue: a quarter of biologists polled by the Federation of American Scientists at the time thought NIH guidelines too lax. Even so, by a second City Council meeting two weeks later, the considerable social authority of Harvard and MIT already began to tell. Gilbert, who had quickly become a local spokesman in favor of recombinant DNA, testified to the lives he would save by cloning the human insulin gene. Instead of Vellucci's proposed two-year freeze, the Council voted for a three-month halt to P3 and P4 work in Cambridge and convened a citizen's review board to study the issue. Harvard and MIT sponsored scientific stars to educate the citizen's board and to debate the few local critics with scientific credibility (particularly Beckwith, King, Harvard's Hubbard, and her Nobelist husband, biochemist George Wald). The City Council ultimately extended the moratorium to seven months, then passed an ordinance that allowed recombinant work to proceed from February 1977, essentially according to NIH guidelines but backed with some additional legal penalties and monitoring provisions. Much the same drama would be played out in Princeton, Berkeley, and other university towns across the nation, as media coverage evoked imagery of Dr. Frankensteins "tinkering with life" and arrogantly unleashing doomsday epidemics. New York and California legislatures also began debating regulatory legislation at state level, while Congressional staff began preparing federal regulatory bills for introduction into Congress in 1977.[18] The Asilomar consensus was emperiled.

At this crucial juncture, biologists joined together as a lobby group to block strict legal regulation of their recombinant work. In late 1976 NIH head Donald Frederickson, the biologist who had already done so much to organize and legitimize the Asilomar consensus embodied in the NIH Guidelines, was struggling to get a blanket Environmental Impact Statement accepted for the NIH Guidelines—in order to save academic scientists from having to prepare a Statement for each individual experiment. Then in January 1977 he caught an alarming glimpse of some of the provisions in the Federal regulatory legislation being drafted by legislators (for example, penalties of up to $50,000 per day of violation, risks that might prevent universities from obtaining insurance). He then summoned a conclave of some 40 eminent scientists in mid-February, including the presidents of the National Academy of Sciences and the American Society of Microbiology and a stellar cast of biomedical professors and deans, and steered them to draft an agreement as to what features a "good bill" should have—that is, what the scientists should push for. Essentially they agreed to join with the drug industry in its new decision to embrace federal legislation, provided it was not too strict and, crucially, provided it preempted local and state laws.[19]

A pivotal moment came in March 1977, when the National Academy convened a meeting in Washington. Unlike Asilomar, this meeting was open to all media and the general public. If its intent was to quell the political waters by confining debate within a narrow band of pro and con positions (embodied in paired speeches by eminent life scientists),[20] it was no success. Decorum was sullied by heckling, not only by unruly protesters from the Left, but also by distinguished scientists from the Right during MIT biologist Jonathan King's comments. (He answered a charge of "Lysenkoism" by replying that the question was not freedom of inquiry but "a question of freedom of manufacture, of modifying the environment.") Activist leader Jeremy Rifkin, allowed a prepared statement to demonstrate the meeting's openness to critics, assaulted the legitimacy of the event as a sham display of democratic concern preceding predetermined outcomes. He also scored points by showing that molecular biology's leaders, like Watson and Joshua Lederberg, had long been contemplating human cloning.[21]

Even the posture that the scientists had the biohazard problem under control wavered under the Klieg lights. For example, when NIH official Dewitt Stetten tried to dismiss insinuations of laxity in testing the safety of tumor virus DNA by reference to elaborate, expensive experiments underway at NIH, an unexpectedly well-prepared activist correctly described Stetten's vaunted P4 facility as merely a trailer behind a chain link fence. But afterward when Stanford biologist Donald Kennedy, the incoming FDA Commissioner, quipped that he could tell "a party is on its way," he was referring to the participating scientists and firms such as Lilly and Dupont, and not the voluble protesters who went home after the opening day of the NAS meeting. There and in other public fora at the time, biologists warded off challenge to their beneficent image by invoking human insulin as the prime example of the medical benefits recombinant DNA would bring.[22]

Foul Play

In San Francisco, Ullrich and Rutter were not as sanguine as Kennedy about the inevitability of permissive regulation. In fact, they were seriously anguished, fearful of what their own success might bring. In late 1976 Chirgwin had devised a new concoction in which to homogenize pancreas tissue that achieved a dramatic decrease in RNAse activity. In his enthusiasm to beat Gilbert, Rutter—Goodman having left for Japan on sabbatical—now asked Ullrich and Chirgwin to work together, organizing a onetime push to extract enough messenger RNA for cloning from the pancreases of some 200 rats, dissected and processed assembly-line style by the entire lab group. The RNA from this batch looked clean, and after enrichment for mRNA and conversion into cDNA, followed by cutting with restriction enzymes and electrophoresis, the sample appeared (somewhat surprisingly) to contain one dominant cDNA. Ullrich treated this cDNA, presumed at 450 bases long to be the main proinsulin transcript, as well as two shorter restriction fragments with enzymes to make them double stranded and blunt at both ends. Then, using a new technique that Boyer and his collaborators at Caltech and City of Hope had recently invented (with input from Gilbert), he ligated a small piece of synthetic DNA containing a restriction enzyme site—a linker—onto the blunt ends of those insulin cDNAs. After running the cDNA-plus-linker fragments on a gel to separate them by size, and then recovering them from the gel band, Ullrich was ready to clone insulin.[23]

But how, exactly? Here technical tractability and the NIH recombinant DNA safety guidelines conflicted. As a cloning vector Ullrich wanted to use a new plasmid from Boyer's group called pBR322. The presence of two separate types of antibiotic resistance in pBR322 made it convenient to determine, first, which bacteria contained the plasmid, and second, which among these carrying inserted genes. In January 1977, the NIH's advisory committee had approved the use of the plasmid, but according to the guidelines, another step based on actual laboratory safety tests—certification—would be necessary before the plasmid could be used in cloning experiments. Evidently confused about pBR322's approved-yet-uncertified status, Rutter allowed Ullrich to use the plasmid.

The cloning procedure was a quick success; by February 1977 the UCSF group had confirmed (through Shine's sequencing) that they had obtained bacteria carrying pBR322 with a roughly 400-base insert, containing almost the whole coding region for rat insulin. To be precise Ullrich obtained several overlapping cDNA clones together carrying a sequence encoding pre-proinsulin, a single long protein 110 amino acids long that is processed by several enzymes in animals into its active, circulating form: a 21-amino-acid A chain and a 30-amino-acid B chain, the two short chains joined by bridges. Ullrich's cloned cDNA contained extra sequences encoding not just the A and B and 35-amino-acid C segment, which in proinsulin connects A and B, but also most of the incompletely determined protein leader region upstream of the B chain which pancreas cells normally remove soon after initiating translation, plus the linkers that Gilbert and Fuller were trying to avoid with their exonuclease/blunt-end approach. And Ullrich's clones were not connected to a promoter either.[24] Still, while they could not produce insulin or indeed anything in bacteria, the rat insulin DNA sequences were enough to claim scientific laurels, both for the technical accomplishment and for the knowledge obtained about the transcribed parts of the insulin gene.

However, the accomplishment could not be trumpeted because it violated the NIH's recombinant DNA guidelines. In February 1977 the California legislature began hearings about whether to follow New York State in giving the NIH guidelines force of law, with stiff fines and possible imprisonment for violations. No wonder Rutter and Ullrich were anguished. Gathering from other scientists that the certification of pBR322 was being significantly delayed (it took until July), in March Rutter telephoned Dewitt Stetten, the NIH official who had clashed with protesters at the NAS symposium only weeks before. Rutter was told that, according to NIH rules, he had to destroy the bacteria carrying the plasmid, and Stetten also expressed great concern about the political fallout if the experiments were reported. But what of the cloned rat DNA? Other researchers who had violated the guidelines had been allowed to first recover and keep purified DNA from their transgressive clones, for in vitro study and transfer later to certified vectors and strains. After consulting an attorney, Rutter decided on this course of action. In registered letters to one another in March 1977, after the conversation with Stetten, Rutter and Goodman bound each other to the agreement that they were destroying the live clones, but keeping recombinant plasmids carrying rat insulin DNA on ice. (They would later testify to the effect that they destroyed the cloned DNA too; see below.) Meanwhile Goodman approached both Genentech and Lilly for "money"—both "for lab" and "consulting"—claiming but only vaguely describing the rat insulin clones that they had obtained. After one of these mid-March phone conversations, Goodman jotted down his discovery that the regulatory violation prevented commercial liaison: "Problem that in Boyer plasmid. Lay low. Not approved. Can't apply for patent yet."[25]

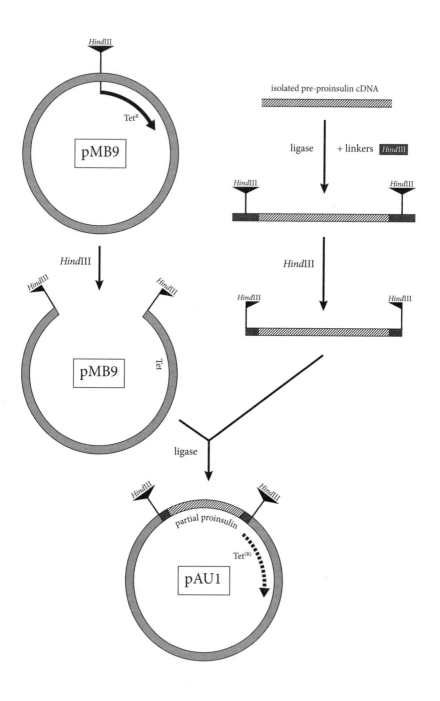

At this point the Goodman-Rutter team evidently attempted, unsuccessfully, to reclone the rat cDNA into the certified vector pCR1. By mid-April 1977, a new possibility emerged when the NIH certified Boyer's vector pMB9, closely related to pBR322. Some three weeks later the UCSF group submitted a report to *Science* describing their recloning of the original rat insulin cDNA into pMB9 (fig. 2.2), and disclosing the sequence thoroughly. On May 23, on word of the report's acceptance, Rutter held a press conference to declare their victory. A court later would decide, however, after review of very extensive evidence, that they had not in fact recloned rat insulin into pMB9 from cDNA, but had used the offending pBR322 clones to obtain the published sequence data, misrepresenting their work in *Science*.[26]

The court also found that this irregular behavior was largely "driven by commercial interests." As Rutter had discovered from consulting the attorney, and Goodman from his efforts to forge a commercial alliance, a patent on an insulin sequence clones in violation of federal guidelines might invite challenge in court—as indeed it did. (And no doubt they were also worried that if the violation were publicized, stricter regulation might follow—a motivation later stressed by Rutter.)[27] Not long after the UCSF announcement of success with pMB9, Ullrich, Shine, and Seeburg learned that the University had applied for broad patents covering general methods for cloning and expressing mammalian hormones in bacteria, including insulin, naming only Rutter, Goodman, and Baxter—the professors—as inventors. As Rutter later conceded, these patent applications "caused a big problem with all of the guys in the laboratories" who did the experiments cited as examples (and, one might add, were needed as witnesses to support his version of events). Ullrich certainly was one who had a problem being left out of the patent.[28]

Fig. 2.2. Construction of the largest rat insulin cDNA clone obtained by Ullrich and other members of the Goodman and Rutter groups at UCSF in 1977, containing a coding region beginning two codons after the first (amino terminal) residue of mature B chain insulin and extending through the remainder of the coding region into the 3′ untranslated region of the gene. Publications describe the use of pMB9 as depicted, with reduced tetracycline resistance due to cDNA insertion in the Tet promoter region, but the courts later determined that pBR322 was actually used to obtain the reported data. (Not all enzymes and substrates depicted; see *Science* 1977;196:1313–19 and main text for further sources and details.) Abbreviations: Tet, tetracycline.

Comes the Dark Horse

The third competitor in this frenetic race for the insulin laurels lagged a year behind Harvard and the UCSF teams at the beginning of 1977. In drafting their first business plan for Genentech in 1976, Boyer and Swanson quickly narrowed their short list of potential protein products, settling on cloning human insulin as their first commercial project (chap. 1). Because insulin was such a known quantity commercially, Swanson was able to fill out the business plan with calculations that the protein could plausibly be produced at a competitive price in genetically engineered bacteria. He also argued that the human hormone would be preferred by patients and doctors (for its lower risk of allergy), and that medical demand would eventually outstrip the pork and beef pancreas supply. Thus, he proposed, the potential market was as large as the $400 million worth of insulin all firms were selling globally by the end of the 1970s.[29]

Swanson's portrayal of the market was far-fetched given the stable preferences of patients and doctors for drug products that reliably did their job. Indeed Lilly had no monopoly on insulin, but its Iletin brand name—exclusively licensed from Banting and Toronto (chap. 1)—had maintained the firm's dominance in the insulin market for five decades precisely through such conservative market momentum. Furthermore, as noted earlier, another technology was emerging to solve the rare problem of allergies to nonhuman insulin: the chemical replacement of amino acids in animal insulins.[30] Still, the "natural" appeal of human insulin, together with its glamorous source in the new gene splicing technology, might give a competitor symbolic leverage to capture Lilly's market. Thus Lilly or one of its major competitors would want to buy genetically engineered bacteria producing human insulin from Genentech. In the widely emulated business plan that Swanson adapted from information technology firms in Silicon Valley, this buyer would pay Genentech in stages, and Genentech would use the revenue to build its own capacities and to develop novel products to sell on its own account. Swanson successfully used this plan to attract venture capital investors, receiving $100,000 from friends at Kleiner-Perkins in mid-1976 in exchange for a quarter of the company.[31] The money was in hand by June, when the first NIH recombinant DNA guidelines came into effect and Cambridge City Hall was about to upset MIT and Harvard's genetic engineering plans.

But unlike their academic competitors, Genentech had little concern about regulatory constraints. Boyer's cloning scheme, unlike Gilbert's and Ullrich's, used synthetic rather than human DNA, and was therefore exempt from the NIH guidelines that were devised to protect human populations from the biohazards of virus genes hiding in animal chromosomes. The two chains of mature insulin were both short enough that DNA encoding them could be assembled fairly conveniently in the test tube by purely chemical methods. After assembly, the DNA sequences encoding the A and B chains of insulin could separately be spliced, using standard enzymes, into plasmids with everything needed to express these synthetic genes in bacteria: a promoter, Shine-Dalgarno sequence, initiation codon, and termination sequences for both translation and transcription. Once the A and B proteins were expressed in separately engineered bacteria and purified, the two chains could be reassembled into mature insulin by (unpatented) published procedures. This synthetic approach allowed tailoring to ensure precisely the correct position for the coding sequence, a big advantage. Gilbert's group, in contrast, was hung up on blunt-end ligating cDNA to the *lacZ* start codon, while Goodman's group was apparently deferring the expression problem entirely.

For the rare capacity to do the crucial DNA synthesis, Boyer had in mind his friend and collaborator Arthur Riggs. Based at the City of Hope cancer hospital near Los Angeles, and collaborating closely with Caltech scientists, Riggs (a Caltech PhD himself) and his colleague Keiichi Itakura had recently synthesized a 21-nucleotide stretch of DNA from the *E. coli lac* operator, which postdoc Herb Heyneker and others in Boyer's group cloned into *E. coli* to demonstrate that it bound *lac* repressor. Indeed, as Boyer knew, Riggs and Itakura had just written an NIH grant proposal to produce a short protein hormone from a synthetic gene in the same way: not insulin but somatostatin, a 14-amino-acid-long protein produced in the brain that counteracts growth hormone. Because the much larger insulin protein seemed a big leap, Boyer persuaded a reluctant Swanson to fund him and Riggs to reach this intermediate goal first. In February 1977, after Swanson had negotiated contracts with City of Hope, Caltech, and the University of California (UCSF) and secured more private financing, the project commenced in these academic labs.[32]

Riggs, Itakura, and Heyneker spent most of 1977 on the project to make the tiny human hormone in bacteria with Genentech funding. (The very fact that Genentech funded a project proposed to NIH without commercial intent shows how much overlap there was at the time between the forefront of molecular biology and the needs, or at least imaginable needs, of the drug industry.) They created a 62-base synthetic gene with sticky restriction sites at both ends: to be precise, an *Eco*RI site followed by a methionine start codon, 42 bases encoding the 14 amino acids of the human protein, two stop codons, and finally a *Bam*HI site. As the methionine start codon implies, this synthetic gene was intended for insertion after a promoter for direct expression in bacteria. But because of second thoughts about bacteria that could churn out growth-retarding hormone in the human gut, the synthetic gene was inserted into the beta-galactosidase (*lacZ*) gene of the *lac* operon, in Boyer's new pBR322 plasmid, so that the bacteria would make a fusion protein from which the hormone could later be freed. After some difficulties that autumn, the bacteria started producing detectable somatostatin. The triumph was announced at an early December, 1977, press conference, where Genentech revealed that somatostatin was a step toward insulin, and published prominently in *Science*. A patent on Riggs' and Itakura's synthetic DNA technique had been filed that, when granted them and exclusively licensed under the terms of the City of Hope funding agreement, would be the first of several giving Genentech a monopoly on this crucial method for producing genetically engineered proteins.[33]

Slow Going at the Universities

Both of the academic groups pursuing human insulin stumbled during late 1977 and early 1978, in part due (in their perception) to NIH restrictions and the threat of still stricter laws. In June 1977, Efstratiadis finally gave Forrest Fuller 1 microgram of insulin cDNA—made from mRNA extracted painstakingly over half a year from the rat tumors and purified from countless formamide gels. In a careless mistake that nearly every graduate student makes once, Fuller promptly lost half in a vacuum drier because he neglected to cover the top of the test tube. Gilbert evidently gave up on direct expression in a single step at this point, since Fuller then attached linkers to the remainder and took the cDNA to Basel for a summer of cloning attempts in Susumu Tonegawa's P3 lab, accompanied part of the time by Gilbert. In August Fuller and Gilbert returned with only 10 transformants (i.e., bacteria carrying foreign DNA), none of which proved to contain insulin DNA.[34]

At the end of 1977 Gilbert effectively dismissed Fuller, who went to another lab to finish his PhD. To replace him as cloner on the Harvard team, Gilbert chose Lydia Villa-Komaroff, a recent PhD from the MIT lab of reverse transcriptase codiscoverer David Baltimore. At the time that Gilbert brought her on board, she was working at Cold Spring Harbor alongside Efstratiadis and Maniatis as a postdoc, cloning human globin (using umbilical cord blood to enable P3 rather than P4, germ warfare-style conditions). Apart from her skill and experience, she was also known as one of the very few people in the United States with a good batch of a key enzyme, terminal transferase, used for the polynucleotide tailing procedure to join DNAs (which Gilbert had previously tried to avoid when he was still trying to get insulin expression in one step, with blunt-end ligation). As a bonus Villa-Komaroff, as an MIT graduate, had little trouble gaining access to that university's P3 laboratory, Harvard's being still under construction.[35]

In contrast with Fuller's efforts over the previous year, and many of her own recent experiments with the silk moth *Polyphemus*, Villa Komaroff's cloning of the rat insulin gene went smoothly and quickly. Using new insulinoma mRNA made by Efstratiadis (but not separated on a gel), she made double stranded cDNA and added a short tail of cytosine "Cs." She then ran the tailed cDNA on a gel and cut out the pieces around 500 bases long, the size expected of a full-length pre-proinsulin message. The extracted cDNA was mixed with pBR322—now officially certified by the NIH—that had been cut open in the middle of the gene for the enzyme responsible for resistance to penicillin, then tailed with guanine Gs. After joining the G-C tailed DNA pieces she induced *E. coli* bacteria to take them in. So far this work only repeated essentially the same steps as the UCSF group, but by using a screening technique involving insulin-specific antibody, the Harvard group recovered bacteria carrying plasmids actually producing insulin protein. These were fusion proteins, consisting of rat proinsulin joined to much of the bacterial penicillinase enzyme—not useful medically, but still the first expression of insulin in bacteria and another scientific milestone. At the beginning of June Villa-Komaroff and Gilbert sent a paper to the quick-publishing *Proceedings of the National Academy of Sciences*. Gilbert also filed patents covering the work on behalf of Harvard which would soon be licensed to the firm he was founding, Biogen.[36]

The group also obtained a human insulinoma tumor, presumably a rich source of human insulin mRNA, and began looking for a place to clone cDNA from it. Gilbert's connections got him a slot at England's P4 germ warfare facility in Porton Down, England, required because DNA from adult human tissue effectively could not be cloned under P3 conditions under NIH guidelines at the time. On September first, Gilbert, Villa Komaroff, Efstratiadis, and graduate student Stephanie Broome (whose specialty was the protein assay) departed from Harvard for their final push to clone human insulin at Porton. As so nicely recreated in Hall's *Invisible Frontiers*, they returned empty-handed from a miserable month of moon suits, airlocks, and formaldehyde footbaths attempting to repeat their rat proinsulin cloning success with the human gene.[37]

Meanwhile, in San Francisco, the postdocs in the Goodman and Rutter labs did more publishable science, some of it impressive. Using the rat proinsulin cDNA he had cloned as a hybridization probe (see chap. 1), Ullrich and the team were able to identify the two distinct insulin-coding genes in chromosomal DNA from rats (humans have only one). They were able to pull one of these insulin genes from gels of restriction-cut rat chromosomes, clone it in smaller pieces, and determine the DNA sequence. They were also able to get a preliminary sense of the structure of the second insulin gene. This work allowed them to address two of the most important fundamental scientific questions in molecular biology at the time: how genes might evolve through duplication (also see chap. 3), and how genes of higher organisms were regulated.[38] Because the keys to transcriptional control of gene expression lay nearby but not within gene sequences expressed as mRNA, this question could only be addressed by cloning chromosomal DNA as the UCSF group had just done (whereas cDNA cloning was adequate and possibly superior for merely technological purposes, that is, making proteins in bacteria).

Rutter, like Gilbert, also managed to source a human insulinoma tumor. The group prepared cDNA from it by the procedures that had enabled them to clone rat insulin. Then Ullrich took the cDNA to Strasbourg, France, where between about June and October 1978 he would try to clone the human insulin gene in a P3 lab Lilly established there (comparable work would remain restricted to P4 conditions in the UK and US until January 1979).[39]

Rutter and Goodman themselves were busy with patent applications, and also with a mounting scandal from their earlier breach of the NIH guidelines. At first, reportedly on advice received in March 1977 from NIH's Stetten to destroy all offending bacteria but to keep the infraction quiet (so as not to aid advocates of strict legislation), Rutter did not formally notify the UCSF biosafety committee of what had transpired. But the affair would not stay quiet. In June a journalist who had interned in Boyer's lab the year before published a story in *Smithsonian Magazine* describing her time among the denizens of UCSF's "Strangelovian laboratory" who delighted in "making genetic chimeras." Observing conversation and behavior at UCSF during the mid-1976 Cambridge controversy, she noted that no more than half the lab workers took safety procedures seriously. She also recorded a prophetic exchange between young scientists. One said "controls are holding up vital research; we have enough growth hormone genes upstairs to clone for a year." He was answered by another who urged him to go ahead with the unapproved cloning because "we're not from the US. No one has to know if you go ahead a little early. You can repeat the experiments later and publish [more quickly]."[40]

That same month, having heard rumours, the UCSF biosafety committee was finally presented with a partial report on the infraction. Perhaps because of that meeting or the *Smithsonian* article, someone at UCSF felt uncomfortable enough to leak the pBR322 infraction to reporter Nicholas Wade, who broke the story in *Science* in late September 1977. Although Wade's article several times asserted that no actual hazard to the public had resulted from use of the uncertified plasmid, and attributed it to honest confusion together with scientific competitiveness, news of the breach raised alarms in a Capitol still embroiled in initiatives to legislate recombinant DNA. The NIH then initiated an investigation, in which Rutter and Goodman were said have destroyed all offending bacteria and DNA back in March; the NIH forgave them. But in November, Rutter and Boyer were called to testify before Adlai Stevenson III's (D-IL) science subcommittee of the Senate Committee on Commerce, Science, and Transportation. By this stage Edward Kennedy, who was originally the driver of legislation from his seat on the Senate's Health Committee, had faded into the background because he did not support federal laws if they preempted stricter local laws. Industry lobbying, with the scientific establishment's help, had made preemption a key feature of any legislation likely to pass.[41]

The antiregulatory tilt of the Stevenson hearings was clear from the first witnesses, Paul Berg and NAS President Phil Handler. Both offered credentials as responsible scientists who had played early roles in airing safety concerns and who—the point already needed to be underscored—had no commercial interests in exploiting recombinant DNA. Both crafted statements that ambiguously opposed legislation (but were consistent with mild regulations), and framed the primary issue as freedom of scientific inquiry and freedom of thought. Berg leaked news of the City of Hope–Genentech success producing somatostatin in bacteria—the manuscript having been submitted to *Science* that very same day—as heralding a cornucopia of new medicines. The Senators treated both witnesses with utmost respect, helpfully asking whether safety guidelines that were too strict might not be taken as seriously as those based in what scientists saw as real dangers. On the second day of hearings a week later, NIH director Frederickson was treated with similar kindness. He suggested that "peer pressure" among scientists remained the best enforcement mechanism. Pressed a little on why UCSF should not have its NIH funding withdrawn for the pBR322 violation, Frederickson replied that his agency's investigation had convinced him that Rutter made an honest mistake, and the offending clones had all been destroyed.[42]

As Rutter and Boyer awaited their turns, Stevenson asked Frederickson if he wasn't "concerned that a policy which permits a university to retain Federal funds after a violation and to obtain and license . . . patents . . . may encourage a lot of innocent mistakes." Rutter in his opening statement gave his version of events, explaining that in March 1977 the UCSF group could simply have reported their insulin cloning violation with pBR322 to NIH formally and thereby "indirectly secured the supremacy of our cloning experiments." (The use of "supremacy" here suggests the intensity of the competition for scientific honor in the insulin cloning race, while the notion that seeking punishment might be a viable way to secure it suggests the degree to which new ways of obtaining credit were being explored.) Instead, continued Rutter, they chose not to out of concern for "the inflamed social and political climate," quietly ceasing work with the plasmids and destroying the violating materials.[43]

Rutter's argument that his behavior was altruistic did not convince some of the Senators. They pointed to evidence that the UCSF group knew pBR322 was uncertified by the beginning of February 1977; that the biosafety logbook of the UCSF P3 lab had been tampered with; and that scientists in his department showed general disregard for the NIH guidelines. The Senators even focused attention on the patentability of the UCSF insulin clones obtained in violation of NIH guidelines. Repeatedly probed, Rutter repeatedly answered ambiguously that he had "destroyed the clones" and that after March 1977 the offending vector was not used. He never stated outright that all the offending DNA had also been destroyed. Still, pressed on the point, Rutter denied that the insulin cloning as described to the Patent Office and published was "based on the research conducted in pBR322 in any way." (This claim the Federal District Court would later discredit, as noted above, finding that the *Science* article and patent applications were based on the pBR322 clones, and that in this testimony "Rutter was not candid" with the Senate or Patent Office.) Stevenson finally dismissed Rutter and Boyer, who felt a sacrificial victim, with a curse: "If there is legislation, it will be because of scientists like you." But with the strongest advocates for strict federal legislation dropping the issue, and both scientific and industrial supporters of genetic engineering seeing no pressing need, legislative initiatives stalled in early 1978. No bill was ever passed to give NIH guidelines force of law.[44]

Genentech's Finishing Dash

When Riggs, Itakura, and Genentech announced their success expressing the synthetic somatostatin gene at a December 1, 1977, press conference at City of Hope, the firm also announced it was cloning insulin. The announcements and acclaim did not convince Lilly to sign a contract with Genentech, as Swanson had hoped. To the big drug firm, established insulin researcher Rutter had seemed a better bet; Lilly instead signed a research contract with UCSF providing his department $250,000 over five years in exchange for exclusive patent rights and know-how, not unlike the sort of arrangement prevalent among drug firms and biologists in the golden age of endocrinology (chap. 1). Still, the somatostatin success was useful for the young company. Swanson raised a third round of private capital and began preparing to outfit a small section of a large warehouse-like building as an office and a small lab. After one-and-a-half years the firm was moving beyond the strictly virtual stage, during which all its work was conducted under contract with academic personnel at academic research facilities.[45]

Swanson now needed to hire biologists to do the cloning, and he looked to the universities near San Francisco. Ullrich expressed interest, but was not ready to leave the academic career path. In December Swanson convinced Dennis Kleid, another young scientist invited through Boyer, to join. Kleid was working in a small lab at the private Stanford Research Institute on soft money from his own grants. Like his even younger postdoc David Goeddel, Kleid's academic training—recent postdoctoral stints in both Gobind Khorana's MIT lab and Ptashne's Harvard group in Cambridge—made him one of the world's few experts on synthetic DNA, the same field as Riggs and Itakura. Small, well-defined segments of synthetic DNA were essential for the precision study of DNA's interactions with other molecules, such as the cancer drugs and carcinogens Kleid's NIH grants funded him to study. But Kleid was tired of grant applications, and had already lowered his academic career ambitions in order to remain in California. He agreed to join Genentech if Goeddel could come with him. The two started around February 1978, with Kleid still spending a lot of time at SRI to wrap up old projects while Goeddel and Heyneker, who soon after joined the firm, helped establish the Genentech lab.[46]

From February through May 1978, the Genentech scientists still did their actual experimental work mainly at City of Hope. Itakura's team had been busy producing the 29 short pieces of synthetic DNA (oligonculeotides or simply oligos), each 10 to 15 bases long, making up the separate A and B insulin chains. Just like the synthetic somatostatin gene, both the 77-base insulin A chain and the 104-base insulin B chain genes they were constructing consisted of an *Eco*RI site, a methionine start codon, then codons specifying the 21 or 30 amino acids (respectively) of the insulin proteins, followed by two stop codons and a *Bam*HI site. Also just like somatostatin, these were to be cloned into pBR322 at an *Eco*RI site near the end of the beta-galactosidase (*lacZ*) gene, where they could be expressed from the *lac* promoter, and then split off from the fusion protein by cyanogen bromide degradation (which breaks proteins at methionines, and fortunately neither A nor B insulin chains contained methionines).

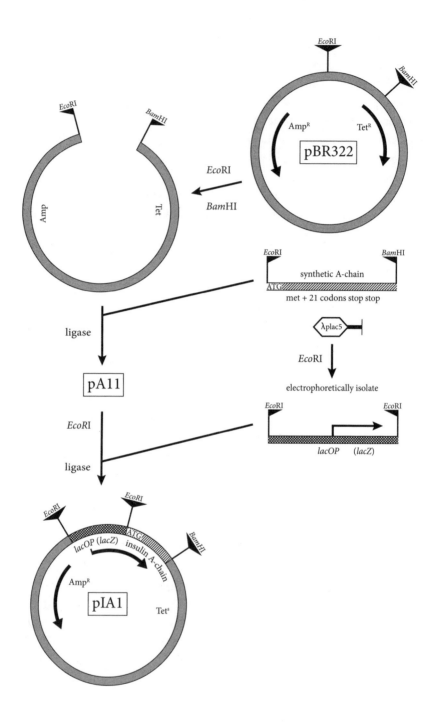

But before the two synthetic genes could be cloned they had to be built up from the 29 oligos. This task had at first fallen to Paco Bolivar, another postdoc in Boyer's group, but he had only partially completed cloning the longer B gene before he returned to Mexico. Taking his place Goeddel—generally with Kleid, too—made regular visits to Southern California for about a week each, during which he assembled and stitched the tiny DNA segments together with ligase in a corner of Rigg's lab. The work days were long and the pace "crazy"—even according to Itakura, who regularly worked 12-hour days himself. Goeddel and Kleid soon finished assembling the synthetic A gene and cloned it in San Francisco without trouble; the second B fragment took longer due to a synthesis error. By September Goeddel and Kleid had recloned the synthetic genes and achieved expression of both A and B fusion proteins from *lac* promoters (fig. 2.3). They began extracting the insulin from cultures by cyanogen treatment, eventually accumulating enough for purification and reassembly into two-chain insulin.[47] Genentech had won the cloning race.

For months Swanson had been in discussion with Lilly, but the big firm resisted commitment, having already put its money behind Rutter. Swanson decided to force Lilly's hand by inviting the firm to the press conference announcing the insulin success. Lilly signed a contract at the eleventh hour so they could share the stage. After giving Riggs and Itakura a chance to bow to the convention of presenting first to a scientific audience (a seminar at Caltech), the press conference took place on September 6, 1978, attracting so many cameras and lights that the power failed before Swanson could speak. But this was no public relations disaster. Press coverage of this "vast leap forward" was glowing,

---

Fig. 2.3. One of two expression constructs produced by Genentech to make the A chain (pictured) and B chain of insulin in bacteria in 1978. The entire coding region corresponding to the 21 amino acid mature insulin A chain was made from synthetic DNA, plus an ATG translation initiation codon at one end and two translation stop codons and the other, and restriction sites at both termini. This synthetic segment was linked in a pBR322 vector to an *E. coli* lactose operator-promoter (*lacOP*) segment obtained from a lambda phage strain carrying the *lac* operon, such that the insulin forms the carboxy-terminal portion of a chimeric protein with a portion of *lacZ*. The insulin protein produced in this way is later freed from the *lacZ* portion of the chimera by cyanogen bromide cleavage, and then purified and combined with the other chain of insulin for reassembly in vitro. (Not all enzymes and substrates depicted; see main text for sources and details.) Abbreviations: Amp, ampicillin; met, methionine; Tet, tetracycline.

and must have pleased Swanson with its reach to the *Wall Street Journal* and the front page of the *Washington Post*.[48]

Soon after he heard the news Ullrich left Strasbourg, his mind made up to join Genentech. The Gilbert group, still working behind airlocks at Porton Down when the news came, gave up their insulin cloning effort. The Rutter group continued and published several more high-profile articles on the natural insulin gene and its expression. The fact that wider biological audiences, like the public, unambiguously declared Genentech victorious implies that, of the two main aspects of scientific progress at stake in the insulin race, advancing the technical frontier of genetic manipulation of bacteria took precedence over advancing knowledge of gene expression in higher organisms. Not that producing insulin from a synthetic gene was anything less than a first-rate accomplishment in molecular biology. However, Genentech won the race by bypassing the natural gene entirely and therefore, unlike Rutter's UCSF group, learned nothing about the insulin gene or eukaryotic gene expression. Perhaps this technical emphasis represents an example of popular acclaim and commercial interest penetrating science's own system of intellectual credit. Perhaps it only reflects a premium on certifiably safe achievements, amid worries still surrounding recombinant DNA's public acceptance.[49]

After the Cheering

The race for the main prize in scientific credit may have finished with Genentech's cloning victory, but from the pharmaceutical perspective the finish line—marketable insulin from engineered bacteria—still lay far away. Swanson's agreement with Lilly was designed to finance the growth of the firm through a series of progress payments, meted out at certain milestones toward commercial production. But to some at Genentech, this agreement seemed a poisoned chalice, because the contract was demanding and stipulated that if scheduled benchmarks were not met, Lilly would be entitled to use Genentech's material and intellectual property without further payment. The second yield benchmark seemed completely impossible to Kleid in September 1978, when he calculated that the weight of engineered bacteria would have to be roughly half human insulin! But scientific ingenuity derived from the best academic biology labs got Genentech clear of the hurdle with which, some at the small firm suspected, Lilly planned to trip them. Giuseppe Miozarri, trained as a postdoc in Charles Yanofsky's Stanford group (the world capital for study of the *trp* operon), swapped the *lac* promoter for the higher-output *trp* promoter, stripped of all the natural control elements that modulate gene expression to meet the cell's need for tryptophan. Linked to a newly re-engineered synthetic proinsulin gene (containing both A and B segments), the cells virtually choked themselves to death with insulin protein when grown in conditions activating the *trp* promoter. Much of the insulin was insoluble, packed into refractile bodies visible under the microscope, but Genentech protein chemists developed methods to coax the proteins back into solution for refolding.[50]

As Genentech's pharmaceutical partner, Lilly took over development of production techniques and steered the product through the many regulatory steps needed to bring a pharmaceutical to market. These included the usual steps required by the FDA: preparing a standard formulation and showing it to be pure and stable; animal testing for toxicity and biological activity; then testing in humans through several phases of clinical trials. There were some novel regulatory issues that would have to be addressed as the new drug moved forward to market, mostly around assuring a safe and reliable drug from bacterial production. Jere Goyan, a pharmacology professor from UCSF who replaced Kennedy as FDA head near the end of the Carter Administration, announced in April 1980 that his agency would proactively convene a special symposium of government and industry scientists to anticipate and resolve any problems. His opening comments at the early June meeting, which attracted representatives of 20 drug firms, made it clear that the FDA was effectively fast-tracking recombinant DNA products. This move was meant to counter drug industry criticisms of over-regulation and perhaps also to be seen as facilitating the economic salvation that the United States anticipated from biotech. True to Goyan's word, FDA and Lilly agreed in advance on novel quality assurance procedures.[51]

By late 1981 Lilly's clinical trials with human insulin were already nearing completion, as were Lilly's manufacturing facilities. In October 1982 the FDA approved Lilly's recombinant Humulin product for sale "in record time," only five months after Lilly applied, reiterating the point very publicly in these early Reagan years that the agency "did not intend to place Draconian regulatory impediments in the path" of recombinant DNA enterprise. As the *Wall Street Journal* observed, at double the price of ordinary insulin and no advantage to most diabetics, the new recombinant human insulin excited doctors less than it inspired stockbrokers, but it eventually would become the standard form of the drug.[52] And of course Genentech's success would soon inspire the formation of many other biotechnology firms, emulating the business plan and company culture.

Making Policy while Doing Science

Genentech did much to set the pattern for the way entrepreneurial biologists would pursue projects that were at once scientifically creditable and commercially saleable in the 1980s—just outside the university's border, rather than within it, the older model from the first endocrinology gold rush on which Rutter tried to build. But the projects themselves in these early days of recombinant DNA generally came from within molecular biology. The race to clone human insulin certainly did, as an effort to demonstrate prowess and win scientific glory by being the first to express a human gene—biology's "most famous protein"—in genetically manipulated bacteria. Counting only the characters mentioned in this chapter, five biologists independently formulated and began concrete projects to clone this particular protein for expression in bacteria, all at approximately the same moment around the beginning of 1976: Gilbert, Rutter, Ullrich, Lomedico, and Boyer. Of these, Boyer (with Swanson's help) most clearly gave careful thought about who might pay for the work, other than public agencies like NIH.

This was no project imposed on scientists by drug firms, venture capitalists, or any other agency diverting science with money. For these scientists and probably many more beyond our narrative, the intellectual values and technical state of molecular biology made cloning a human hormone into an bacteria obvious and irresistible next step after the Cohen-Boyer experiment. The leaders among them, and their fellow senior molecular biologists throughout the world, appreciated the attraction and wanted the path smoothed for more such biology, choosing insulin as their emblem and opening move. And the biologists broke the old conventions of apolitical behavior, reshaping society to fit these ambitions, especially in the United States.

Biologists defined and then addressed their own safety worries through the Asilomar consensus-forming process, in an effort both to prevent public harm and to preempt regulation. Afterward they forestalled strict regulation by quelling public fears, individually writing in the popular press and speaking in forums big and small, and by intervening in legislative debate via collective bodies such at the National Academy of Sciences. If we can count NIH chief Donald Frederickson and FDA heads Kennedy and Goyan as biologists (the latter two were life science professors on furlough, the first a medical researcher who moved from Harvard Medical School to NIH early in his career), then we can also say that biologists in government service made sure that red tape such as environmental impact statements and pharmaceutical quality standards would not obstruct genetic engineering in the laboratory or the marketplace.

Leading academic biologists like Rutter and Gilbert also enrolled their own universities in the new game of funding molecular biology through commercial linkages, although this began haphazardly and would take time. Legal changes making it easier for scientists and universities to sell their work as intellectual property would speed and spread the shift, and the biologists helped make these changes happen too. One key early moment came in *Diamond v. Chakrabarty,* a case that concerned the patentability of a bacterium engineered (by nonrecombinant means) to break down oil spills. In 1980 the solicitor general, the federal government's chief legal official, appealed this case to the US Supreme Court to resolve whether genetically engineered organisms could be patented under existing law.[53] The Supreme Court received ten *amicus curiae* briefs, with only one opposing patenting—and that from Rifkin, who argued that commercial genetic engineering would not benefit the public. More interesting are differences between the other briefs. The Pharmaceutical Manufacturers Association supported patenting not because it was essential for the commercial exploitation of recombinant DNA, but because without patent protection the industry would rely on trade secrets that would retard progress in the long run. Many in the drug industry and on Wall Street expected patents to be comparatively unimportant in genetic engineering anyway, since they would become too quickly obsolete.[54]

In contrast stands the brief representing a large part of the American biomedical research community, submitted by the American Society of Biological Chemists, the Association of American Medical Colleges, the American Council on Education (another university lobby group), and several scientists including Tom Maniatis and Leroy Hood at Caltech. Unmentioned in the brief, these last two were already involved in starting the biotech firms Amgen and Genetics Institute respectively (chap. 5), with no plans to leave their university posts. Even "though engaged primarily in basic research," all these parties claimed

> an interest in seeing their work reach commercial development. They fear that adoption of a *per se* rule excluding all living things from patentability will inhibit commercial development of the advances they are making in recombinant DNA research. . . . For example, it is a scientific breakthrough to clone the interferon gene in a bacterium. However, to maximize the production of interferon from such a modified bacterium will require a great deal of additional scientific expertise which will not be forthcoming without the protection and rewards that

patents provide to the scientific investigator and to the commercial sources which normally fund such work.[55]

Given that, as industry well understood, commercial drug development through genetic engineering would continue with or without patents, what Hood, Maniatis, and Caltech were really talking about was whether academic biologists would enjoy "the protection and rewards that patents provide" for doing genetic engineering—rewards which would otherwise go only to drug companies. In another *amicus* brief the American Society of Microbiology, "the largest single life science organization in the United States," echoed this position that patents were needed to reward (university) scientists for commercially oriented genetic engineering. Given the Court's close 5–4 decision favoring patent protection for genetically engineered microbes as socially useful and not especially dangerous, it is reasonable to suppose that the seemingly unanimous voice of the American biomedical research community played a role in the outcome, particularly since the business community's voice on the issue was equivocal.[56]

Similarly the 1980 Bayh-Dole bill, which encouraged universities to patent and commercialise work funded by the NIH and other federal government agencies, was an initiative that came largely from science-driven universities themselves, particularly from those that had built up technology transfer offices during the 1970s.[57] Even if no scientists testified at the Congressional hearing preceding its passage, patent-holding academic leaders of biotechnology like Gilbert and Rutter could hardly object, and apart from a small activist minority like those associated with Science for the People, very few life scientists raised worries about university patents at the time. Some, like David Baltimore, argued that patent ownership actually would give academic scientists more power to negotiate and retain independence from industrial funders.[58] As we will see, an attitude prevailed among American life scientists that if anyone should get rich from exploiting NIH-funded research, it should be the publicly funded scientist who had done most of the relevant work. Academic biologists moreover played a central role in cultivating an investment market for biotechnology. In all these social roles, life scientists fostered and shaped the biotechnology sector as it emerged according to their values and interests, just as they had created the NIH recombinant DNA guidelines.

Sociologist Sheldon Krimsky, a member of the Cambridge City Council's citizen study group, provides insight into how the biologists won this and other battles of the late 1970s and early 1980s. Whereas critics of recombinant DNA research offered a wide range of reasons and opinions to the City Council body, the mainstream supporters all "presented a remarkably united front," endorsing the NIH guidelines as sensible and adequate. This unity in 1976 stands in sharp contrast to the differences of opinion the biologists aired among themselves in 1975 when constructing the NIH guidelines. In the face of damaging, sensationalist press coverage and political initiatives threatening unprecedented legal controls, the molecular biologists closed ranks around the NIH guidelines forged by Berg through the Asilomar self-regulation process. And the single-voice strategy ultimately worked, in that the few local ordinances actually imposed did little but give the NIH guidelines force of law, and federal legislation was never passed in the United States. But where national laws did pass, such as the Britain's occupational health and safety law, they also conformed to the NIH guidelines. From this perspective Berg's Nobel prize must be regarded as an award not just for scientific innovation but also political leadership: his Moses-like role leading the molecular biologists through the desert of public scrutiny, to an abundant land where genetic engineering within their research institutions could profit them and the outside firms they started.[59] That Berg himself would never spend much time in the promised land of biotech only strengthens the analogy.

CHAPTER THREE

Growing Pains

Commercial Strains on a Way of Life

In the shelves full of books written about the biotechnology industry and its meteoric rise around 1980, tales of heroism abound. Exceptional qualities have been attributed to the biologists who first cloned human proteins to make new drugs possible: foresight and genius, brilliant entrepreneurial creativity, dedication to inventing lifesaving new medicines, even the same idealistic ambitions to better the world that typified youthful radicalism of the 1960s. A counternarrative of corruption is also common, as greed stimulated by policy (for example the 1980 Bayh-Dole Act, encouraging University patenting) diverted molecular biologists from their noble and socially useful quest for basic knowledge into get-rich-quick projects with commercial promise but little scientific value, undermining ethics along the way. From all sides, revolutionary implications for both biology and the pharmaceutical industry have been pronounced, especially centered on changes in the ethos and intellectual orientation of the molecular biologists who entered the new arena of commercial molecular genetics.[1]

In this chapter we will use the lens of human growth hormone, one of the first biotech drugs to reach the market and the basis of Genentech's initial growth, to look at how scientific work and professional behavior shifted among the first molecular biologists entering the commercial arena. We will see that the shift was not radical. Much as in chapter 2, where I argued that the first recombinant DNA product to be commercialized as a drug was bound to be insulin—so dictated by the history, values, and technical capacities of academic molecular biology in the middle 1970s—this chapter explores more thoroughly the way that the general project of cloning protein hormones was determined by molecular biologists following established research trajectories in the field, and expecting a payoff in scientific credit for their accomplishments.

Biologists both inside and outside the commercial sector drew up much the same short list of proteins that could be cloned; as the president of the Cetus biotech firm grumbled in 1979, "everybody has the same damn list." Many lists were topped by insulin, but growth hormone appeared on more than one as well. The situation is rather like the simultaneous scientific breakthroughs that have long fascinated historians, for example Newton and Leibniz's independent discovery of calculus, or the formulation of the law of conservation of energy around 1845 by at least four different scientists (Joule, Thomson, Colding, and Helmholz). The synchronous movement of multiple scientists suggests both a common thought process and a context affecting them all in the much same way.[2] Biologists responding early were unusual only in their sensitivity to opportunity and incentive. Nor is any exceptional dedication to saving lives evident; apart from interferon (chap. 4), none of the early drug development targets adopted by the first generation of genetic engineers held particularly great medical promise at the time. Indeed, there was absolutely no demonstrable need or demand for recombinant human growth hormone when the effort to clone this protein began. Thus there is no room for inexorable market forces, no possibility of economic determinism in explaining it. Like the goal, the specific path that the cloning effort took came from the internal logic of molecular biology, as filtered by local resources and priorities. As with the other cloning races of the late 1970s, biologists pursuing growth hormone were just scientists trying to do science, driven by at least as much by the traditional contest for standing in their field as any remote and questionable possibility of becoming rich.

A Logical Project

Growth hormone, or somatotropin, presented itself as an obvious cloning target for early genetic engineers for a number of reasons. Though not a thoroughly familiar object like insulin, the substance was rather well understood scientifically by the mid-1970s, thanks to more than three decades of basic research. The growth-promoting protein hormone made by the tiny pituitary gland had been discovered in the original endocrinology gold rush, between the world wars. During the 1940s, Berkeley endocrinologists Herbert Evans and C. H. Li emerged as leaders in the biochemical analysis of the hormone, extracting it from slaughterhouse ox brains, tracking it by bioassay, and in 1948 achieving crystalline purity. Li, still active in the mid-1970s, was nominally a member of Rutter's UCSF Biochemistry department. He had made a lifetime's career of studying the protein according to the pattern of mid-century endocrinology, gaining several patents on it, and building up a dedicated research institute around himself.[3]

The hormone's sequence was almost entirely available in the published literature in the mid-1970s, thanks in part to corrections to Li's work by other protein chemists. Two other protein hormones in the same family were also known, both with functions related to pregnancy: lactogen, made by the pituitary gland, and somatomammotropin, made by the placenta. The roughly 200-amino-acid size of growth hormone (or 600 bases, coded in nucleic acids) made it tractable for isolation and cloning, by the techniques of the day. The good luck that natural human growth hormone was pure protein, lacking any of the exotic sugars that are often added by higher organisms in producing other large proteins, also made it attractive for ambitious genetic engineers with drugs in mind. How growth hormone affected metabolism and growth in various tissues was also understood fairly well (certainly better than many hormones), as well as its pharmacology in animals and humans.[4] From a life scientist's point of view, then, human growth hormone was a known quantity presenting no special obstacles for cloning and expression in bacteria.

Somatotropin was also a known quantity from the medical perspective. Its usefulness for treating children with retarded growth due to pituitary deficiency, a rare condition, had been proven in multiple clinical trials since 1950. But because humans only respond to the protein from humans and apes—not from cows or sheep or pigs—it could not be commercially manufactured from slaughterhouse sources, like insulin and many other hormone drugs. But the hormone did become commercially available around 1970, when the supply problem was overcome through international networks collecting human brains from recently deceased donors, from which the pituitary protein could be extracted. The world's largest producer of human growth hormone in the middle 1970s was not a drug leviathan like Merck or Lilly but the specialist government-owned Swedish firm Kabi, which produced enough from hundreds of thousands of brains to treat about 3,000 cases of pituitary dwarfism per year globally, at a price of $5,000–10,000 each. While some Kabi executives felt that supply was limiting sales, a demand for more or better growth hormone was not obvious to many close to the situation. As late as 1979 the head of the US National Pituitary Agency (NPA), a nonprofit that served most of American medicine, asserted that "all patients with diagnosed hyposomatropism are receiving the hormone" in adequate dosage from cadaver pituitaries.[5]

Growth hormone is now remembered as Genentech's first product. As we shall see, when Genentech decided to take it on, financed by a contract from Kabi, the effort to clone human growth hormone had been underway at nearby UCSF for two years and was well advanced. This project began in academic biology, for academic reasons. Beyond technical plausibility, the prospect of cloning human growth hormone offered important intellectual rewards to molecular biologists in the mid-1970s. As noted above, biochemical and immunological research had shown that growth hormone belonged to a gene family known to share much of the same protein sequence. According to an exciting and then-new theory, the complexity of higher organisms evolved through the accidental duplication of genes, which then took on different forms and functions as their expression was taken over by specific tissues at different times in development. The closely related genes of the growth hormone family (including lactogen and what tuned out to be several somatomammotropin genes) presented a clear-cut example, so that the process of gene duplication and divergence could be studied by comparing their DNA sequences—which had just become technically more feasible through new methods of DNA sequencing.[6] And if the regulatory sequences situated on the chromosome near the genes could also be cloned and sequenced, they would shed light on the mystery of how higher organisms expressed particular genes with such exquisite control over timing and position within the organism. Thus the cloning of the growth hormone family of genes was a project combining three key fascinations of 1970s molecular biology: developmentally specific gene regulation (the long-sought eukaryotic operon),[7] how genes evolve at the sequence level (the duplication hypothesis), and the sheer technical challenge of cloning and expressing a specific eukaryotic gene in *E. coli.*

Growing Up Quickly

When Peter Seeburg came to San Francisco in mid-1975, fresh from his PhD in Heidelberg with bacteriophage researcher Heinz Schaller and funded by a West German government fellowship, he arrived with a clear—if naive—plan to be the first to clone and express a human hormone in bacteria. Seeburg had an agreement with a leading nucleotide chemist in Germany to supply him with synthetic DNA coding for the very short (8 amino acid) hormone angiotensin-2, which raises blood pressure, for cloning in Boyer's lab. Seeburg's new ambition drew on his doctoral work, which dealt with gene regulation in the bacterial virus M13 and the identification of prokaryotic promoters, mainly in that the new project required understanding of gene expression in bacteria. His motive for cloning angiotensin-2 was strictly the quest for scientific distinction; there was—and still is—no conceivable demand for another injectable drug to raise blood pressure. Frustratingly, shortly after his arrival in California, Seeburg and Boyer's erstwhile collaborator had a change of heart and decided that rather than simply giving him the angiotensin DNA, Seeburg should instead learn the latest techniques and return to Germany to clone it in his own lab. This possessive attitude foreshadowed one that Seeburg and Ullrich would frequently encounter among the senior scientists at UCSF.[8]

So it was that soon after his arrival in San Francisco, Seeburg was left without the hormone-cloning project he had set his heart upon. That autumn of 1975, he chanced to bump into a casual scientific acquaintance in an elevator and mentioned his plight. His fellow traveller was John Baxter, whom Seeburg had first met at a European conference. In 1975 Baxter, a researcher studying regulation of hormone secretion, was working on the hormone responses of cells derived originally from a rat pituitary tumour. This particular cell line not only grew well in tissue culture dishes, but also produced high levels of growth hormone when triggered under the right culture conditions. From a molecular biology perspective, Baxter had a nice system for studying the regulation of growth hormone gene expression, with his responsive cell line, and also a useful assay involving an antibody that specifically bound the growth hormone. Their conversation made further potential obvious to both: the rat tumor line could supply a large enough and pure enough quantity of messenger RNA to convert into cDNA with reverse transcriptase, and once in cDNA form, the coding portions of the growth hormone gene could be sequenced and cloned. Doing this with the rat gene could lead the way to the human hormone, because it presumably had almost the same sequence. With Joe Martial, a Belgian postdoctoral fellow in Baxter's lab, Seeburg thus pursued a new version of his original plan, to clone a hormone—now growth hormone—via the rat tumor.[9]

Around the beginning of 1976 Seeburg launched into this project with Martial and Baxter at night. The three were moonlighting because they felt that Boyer and Rutter and Goodman, into whose lab space Seeburg had been transferred to take up another project, did not approve of Baxter. The aversion was partly generic, reflecting attitudes that (I know from experience) many molecular biologists and biochemists would have held under the circumstances: Baxter had no claim to molecular expertise, and worse still was trained as an MD and not a PhD—thus lacking in rigor and intellectual seriousness. But there was also a more specific animosity toward Baxter: as Biochemistry Department Chair Bill Rutter later recalled, Baxter had been the protégé of his own friend and predecessor as Biochemistry Chair, Gordon Tomkins, but the two had become estranged when Baxter took over research areas that Tomkins had developed. Thus Baxter could be seen as a poacher.[10]

Seeburg also feared that, had they known about it, Boyer and Goodman would have regarded growth hormone as a distraction from a new project that they had set him on after the angiotensin plan fell through. As he later put it, Boyer and Goodman talked him into a project similar to his PhD work, an effort to identify promoter regions of random genes in the fruit fly *Drosophila* by passing fragmented fly chromosomes through a column containing immobilized RNA polymerase. It was a fishing expedition, almost literally—in that he was attempting to catch DNA that would bind the enzyme bait—but also in the metaphoric sense that this work seemed a long shot to Seeburg, much less likely to yield results than his thesis project with the simple phage M13. To be sure, it might have been intellectually cutting-edge and highly rewarding in the unlikely event of success in isolating fly promoters, since the eukaryotic operon was still the Holy Grail of molecular genetics. But to Seeburg it was a disagreeable diversion from his goal to clone a hormone.[11]

Half-heartedly pursuing the fly project by day, Seeburg soon expanded his moonlight work with Martial and the rat pituitary cell line to human hormones. In 1976 Seeburg, Martial, and Shine started extracting RNA from human placentas, easily scavenged from the waste stream of the maternity wards. The target here was the hormone somatomammotropin, one of the main genes expressed in placentas. Eventually the work could no longer continue in secret. When Seeburg finally filled Goodman in on their progress in mid-1976, in order to talk about writing up the first joint publications from this hormone cloning work, Seeburg recalls Goodman's eyes "lighting up" with interest.[12]

With Goodman now part of the project, Seeburg and collaborators sketched a plan for cloning human growth hormone. The first step would simply be to isolate the specific cDNAs of the two genes related to human growth hormone (rat growth hormone from Baxter's tumor cell line and human somatomammotropin). Then these could be cloned into bacteria to enable the complete reading of the sequence using the new Maxam-Gilbert procedure that Shine had mastered. These cloned rat cDNA sequences might in turn help them clone human growth hormone by serving as a probe—that is, exploiting DNA's complementary binding property—to identify matching sequences on a gel or in bacteria carrying cDNA derived from human pituitaries. The same rat somatotropin and human somatomammotropin clones might even be used as a probe to fish the genes out from chromosomal DNA, which unlike cDNA would contain the promoters and nearby sequences sought after for the study of gene regulation in higher organisms. And of course the sequences of the related hormones from the two species could be compared, enabling a look at the evolution of gene families.

All this would yield intellectual rewards and scientific credit, but nothing of direct medical or commercial use. For that an extra step would be required: to use bacteria for making human growth hormone, the cloned DNA sequence for the human hormone would have to be re-cloned into another plasmid, after trimming to remove extraneous DNA sequences, and lined up perfectly so that the bacteria would read it into RNA and then into protein. Unlike Gilbert and Fuller's ill-fated strategy of cloning a shortened insulin cDNA straight onto the *lac* operon without linkers or tailing, the UCSF team showed no advance interest in addressing this problem of expression (just as with their own insulin effort). During Goodman's sabbatical absence in late 1976 and 1977, Baxter and the postdocs simply launched into a direct rush for the scientific prizes to be won by isolating the genes and cloning them into bacteria.

The first major article from Seeburg's growth hormone work was completed in January 1977. It described how Baxters's rat tumor cell line could be made to produce large quantities of growth hormone messenger RNA. That the mRNA largely coded for growth hormone, they verified by extracting RNA and making proteins from it in a cell-free translation system (a test-tube concoction of protein-synthesizing enzymes and factors derived from wheat germ). They also used reverse transcriptase to produce rat growth hormone cDNA from the mRNA they extracted from the pituitary cell line, a source of raw material for later cloning, and perhaps what the unnamed foreign postdocs overheard by the Smithsonian journalist were so eager to proceed with in mid-1976, NIH rules notwithstanding (chap. 2). Three months later, the group sent its first major article on somatomammotropin to a journal. As with the rat tumor, they showed how RNA extracted from human placentas made lots of the hormone when added to the wheat germ system—but also several other proteins in lower abundance, from the mixed set of mRNA produced by the placenta tissue. Adding reverse transcriptase to make cDNA copies of the mixed placenta mRNAs, they cut this diverse cDNA with the restriction enzymes *Hha*I and *Hae*III and separated the resulting DNA fragments by size in an electrophoresis gel. The most abundant discrete bands presumably came from the mRNA for somatomammotropin hormone, placenta's major protein product. The longest (*Hae*III) fragment was about 550 bases, so it almost certainly represented a partial sequence for the hormone messenger RNA, known to be roughly 800 bases long. Shine confirmed these fragments' identity as somatomammotropin by slicing the cDNA fragments from the gel, and determining that their partial base sequence corresponded to the published amino acid sequence of the hormone.[13]

These early publications with the rat pituitary and human placenta hormones represent a dress rehearsal for the cloning of human growth hormone. The team had learned how to pull specific hormone-coding cDNA fragments from a heterogeneous mess of RNA extracted from cells. They had stopped short of cloning these hormone sequences, but by extracting particular cDNA fragments from a gel (which could as an extra precaution be cut again with restriction enzymes, then rejoined), they had generated a "purified" natural gene fragment coding for the hormone, which could be put into bacteria without raising fears of transferring hidden virus genes, and thus with regulatory requirements lower than those associated with shotgun cloning.[14]

Cloning soon followed, building on these technical accomplishments. Two more high-profile papers bearing the names of Seeburg, Martial, and Shine, as well as the professors in theory overseeing their work, Baxter and Goodman, were submitted in September 1977 and published that December, back-to-back in the very eminent journal *Nature*. The authors reaped intellectual reward by describing the sequence of the rat pituitary and human placenta hormones in depth. As before, the team had converted hormone mRNA extracted from tissue into cDNA with reverse transcriptase, cut the heterogeneous cDNA with restriction enzymes, and then isolated the hormone gene sequences by slicing the dominant bands from an electrophoresis gel separating the cut cDNAs by size. These hormone gene fragments from the gel were cloned and thoroughly sequenced by Shine. One *Nature* paper described most of the coding sequences of the rat growth hormone gene, which actually encodes a hormone precursor or prehormone (rather like proinsulin) of 216 amino acids, later processed by special enzymes in the cells that secrete it to a final 190-amino acid form. Using linkers, Seeburg and Martial had cloned a large piece of the hormone and, in a remarkable feat, also cloned an 800-base cDNA corresponding to the entire prehormone messenger RNA, complete with extra sequence before and after the protein-coding section, straight from a faint band on the gel separating uncut cDNAs by size. These noncoding regions provided clues as to the mechanisms of gene translation in eukaryotes.[15]

This tour de force was coupled with an account of how the team had similarly cloned most of the gene for human somatomammotropin by extracting placental mRNA, converting that RNA to cDNA with reverse transcriptase, cutting the cDNA with restriction enzyme *Hae*III, isolating the dominant 550-base band of cDNA that now appeared on a gel separating DNA by size, and splicing it into Boyer's pMB9 vector. Shine sequenced the cloned *Hae*III fragment and found that it specified all but the first 23 amino acids of the mature, 191-amino acid placental hormone. Meanwhile, Baxter started searching for brain tumors through his medical connections. There was every reason to believe that exactly the same steps could be used to clone the closely related human growth hormone, once a pituitary source of mRNA as good as the placentas was found.[16]

Freezer Burn

Through late 1977 and most of 1978 Seeburg had continued as before to work with Martial and Baxter, and nominally Goodman too, although relationships among the principal researchers were starting to fray. A paper submitted to *Nature* in October 1978 and printed that December describes some of their progress, although nothing so dramatic as the massive accomplishments previously published. In brief, the DNA sequence from the cloned full-length rat growth hormone message was shortened slightly and fused to the middle of the gene encoding the enzyme for resistance to the antibiotic ampicillin in Boyer's plasmid pBR322. When put in live *E. coli* the spliced plasmid produced a long chimeric protein, half antibiotic-digesting beta lactamase enzyme and half rat growth hormone (including most of the pregrowth-hormone), rather like what the Gilbert team had accomplished with rat proinsulin about half a year earlier. There was no way to separate all these superfluous amino acids from the growth hormone they were fused to, and even if this were possible the rat hormone had no medical use. Still by achieving expression Rutter, Goodman, and Baxter may have signaled progress toward a drug to the executives and shareholders of Lilly, with whom they had signed a contract to fund UCSF covering not just insulin but also growth hormone (chap. 1 and below).[17]

In fact Seeburg and his UCSF colleagues had accomplished much more in the first half of 1978 than the published literature showed. As revealed by vague references to "unpublished data" in a 1979 paper and more fully in a patent applied for in April 1978 and finally issued in December 1982, Seeburg and the other postdocs had managed to clone the 550-base segment coding for all but the first 23 amino acids of human growth hormone protein. They had done this by extracting mRNA from human pituitary tumors flash-frozen straight from brain surgery, making cDNA and cutting it with *Hae*III, then cloning the band sliced from a gel using linkers. In all this Seeburg, Martial, and Shine followed almost exactly the same steps taken with the comparable fragment from placental somatomammotropin in 1977 and described in *Nature* (fig. 3.1). The gene for the human hormone was then sequenced. The fact that this tremendous accomplishment was not promptly reported in *Science* or *Nature*, unlike the rat and placental hormone results, suggests that by this stage the practical and commercial goal of producing cloned human growth hormone in bacteria had come to influence the traditional quest for credit from publication. Scientific credit was not actually foregone, but postponed; assuming the patent would eventually be granted, priority was secured by the (in effect) secret patent application. But commercially motivated competition meant that claiming the honors in print would have to wait. Of course it mattered less to lab chiefs—their bellies "with good capon lined" (as Shakespeare put the generational difference)—than to postdocs with academic reputations to build, that full public credit be claimed sooner rather than later.[18]

When Rutter committed the University of California to the commercial development of the hormones the postdocs had been cloning, he had presumed too great a claim to credit and to authority over those actually doing the work, at least in the view of these junior scientists. The intergenerational trouble first erupted into open conflict after the May 1977 press conference that UCSF called to announce the cloning of the rat insulin gene, in which the postdocs (especially Ullrich) felt that they were not given proper credit.[19] It was soon after this that Ullrich and Seeburg learned about the University's general patent application for methods for cloning and expressing human hormones—naming only the professors as inventors. The university abandoned that generic application, and eventually reapplied with more specific patent applications naming Seeburg, Ullrich, and other postdocs leading the lab work cited as examples, but not before seriously upsetting them.

The postdocs had immediately perceived that being named as inventors on a patent would confer credit or honor in much the same way as authorship on an important publication (that is, in sociological terms, inventorship would secure symbolic capital as well as potential royalties). Soon patents would become routine parts of biologists' academic resumes. It is as if, once the norm of prompt journal publication wavered, life science quickly reverted to a state prevailing before the modern system of credit-for-publication was established; Galileo, for instance, sought patents to secure credit for inventions without describing them fully, and scientific secrecy was one of the main problems that the Royal Society's ethos of publication-based honor was supposed to alleviate (see below). Certainly, Rutter recalled, his first efforts to file patents that would bring funding to his department in 1977, "caused a big problem with all of the guys in the laboratories" because of the credit implications.[20] Or, as Seeburg recollected more vividly:

[We] were not on the patent and I was really pissed, I was upset. So I still remember, I went to Rutter's office—he was the department chair—and I said, "This patent contains . . . nothing in it that you did or designed or wanted to do."[21]

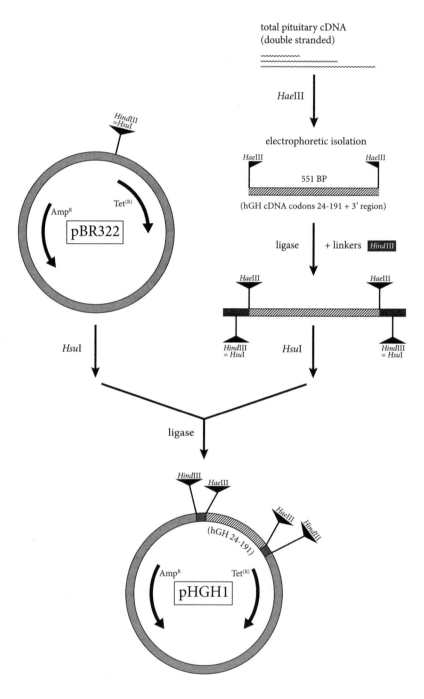

Rutter's reply made him fearful for his career prospects. Unsatisfied with his place at UCSF, Seeburg applied for other academic postdoctoral positions late in 1977, and eventually received an offer from an East Coast lab. He also thought about joining Genentech, where (by early 1978) his UCSF friend Heyneker was already working. The contract that Rutter had meanwhile negotiated with Lilly would bring the Biochemistry department $1.25 million over five years, in exchange for the patent right to insulin and growth hormone genes cloned at UCSF. The professors had sold Seeburg's research project without including him in the bargain.[22]

About one week after the September 1978 press conference where Genentech announced the successful cloning and expression of their synthetic human insulin genes, Genentech convened a small meeting in San Francisco with the disaffected UCSF postdocs to plan their attack on growth hormone. Genentech had just signed a contract with Kabi to develop recombinant human growth hormone, the Swedish firm having approached Boyer and Swanson on the heels of their somatostatin success in December 1977. Like the insulin contract with Lilly and the urokinase contract with Gruenenthal (chap. 6), Genentech's August 1978 growth hormone contract with Kabi was designed so that the sponsor would pay in installments when the small firm met certain benchmarks, and later pay royalties on sales. Kabi would have exclusive rights to sell the product except in North America, where Genentech would share the rights.[23] Swanson's agreement to share such a small market, the roughly 2,500 Americans served by the NPS, implies that he had no inkling in 1978 of Genentech actually marketing the hormone.

To get paid, Genentech had to work fast, and the UCSF postdocs could help them. According to Dennis Kleid's vivid recollections, this September meeting brought Swanson and himself together with Itakura and Roberto Crea from City of Hope, Heyneker (just returned from Holland), and Seeburg, Ullrich, and Shine. Kleid and Goeddel had decided beforehand that DNA encoding the entire 191-amino acid growth hormone was too long to generate chemically (as they

---

Fig. 3.1. Unpublished cloning of the bulk of the human growth hormone gene's coding region accomplished by Peter Seeburg and others associated with Howard Goodman's UCSF group during 1978. The cloned *Hae*III cDNA fragment encodes amino acid 24 of mature growth hormone through the remainder of the protein, as well as 3′ termination sequences. (Not all enzymes and substrates depicted; see US Patent 4,363,877 and main text for further sources and details.) Abbreviations: Amp, ampicillin; hGH, human growth hormone; Tet, tetracycline.

had with insulin). On the other hand a cDNA gene made from growth hormone mRNA would be hard to trim to eliminate the leader section and line up perfectly with the promoter driving transcription in an expression vector. A better way would be to build the first part of the growth hormone gene synthetically, including the methionyl start codon and the beginning part of the hormone up to an early restriction site and then, from that restriction site onward, splice in cDNA obtained from natural mRNA. From the meeting emerged a detailed plan to synthesize the gene from the promoter up to amino acid 23, and then for the 24th amino acid onward insert the 550 base *Hae*III fragment that Seeburg had recently cloned.[24]

The three postdocs wanted to stay together, like three musketeers, so they all agreed to join Genentech and the relative freedom and fair reward it promised. Ullrich must have wavered, however, because he then went off to France to clone proinsulin in Lilly's facility as part of the UCSF team; he later recalled committing to Genentech only in October. Shine soon returned to Australia for an academic post. Seeburg was more committed to the Genentech project, still unhappy with his relationships with Goodman and Rutter at UCSF but having turned down the East Coast opportunity. However, Genentech would not have lab space ready for him and Ullrich to work in until January 1979, when they would both officially become employees. So he continued doing experimental work at UCSF facilities in late 1978, largely on what he described later as his own scientific project, concerning the evolutionary divergence of the somatotropin gene family.[25]

One day in November 1978, when Seeburg recalls that he was showing Goodman the results of sequencing of the growth hormone-related clones, Goodman abruptly ordered him to drop the project—which he had now been working on for three years. Goodman moreover banned him from the lab, effective immediately. When Seeburg returned to UCSF later that evening to empty his desk and retrieve samples of his DNA and bacterial strains, he found that Goodman had taken all of his materials out of his freezer and locked them in another. Obtaining keys from another scientist, he found that all the tubes from his freezer had been dumped from their racks into a single large pile. Given that most were not labeled and could only be identified by their positions on the original racks (as was typical in fast-moving, unfinished projects), this meant that many months of work was lost, much of which had nothing directly to do with human growth hormone. A plausible rationale for Goodman's destructive act was to prevent Seeburg from taking his human growth hormone clone elsewhere—assuming that Seeburg did not keep back-up tubes of his most important work elsewhere. Seeburg had made no secret that he would take with him samples of the growth hormone recombinant bacterial strains and other products of his work at UCSF, as was traditionally a postdoc's right when moving from one academic lab to another.[26] Whether that traditional journeyman's right to his work products extended to a postdoc moving from an academic lab into a biotech firm was a crucial question, especially when the academic institution had (just like a private firm) contracted with a drug company to provide exclusive rights to those same work products.

But perhaps, in addition to a rational if violent claim to intellectual property, we might also read an element of emotionality into the freezer incident. For it was around this time that negotiations to bring Goodman into the Genentech fold had broken down.[27] Whatever the motives, this event marks a battle line drawn in the conflict between the professors and the junior scientists over control of both the knowledge and the materials being produced at UCSF. Rutter and Ullrich severed relations only a little later, albeit not so abruptly.

Triumph of the Rebellion

Seeburg and Ullrich retaliated with a famed midnight raid, in which they visited the old labs at UCSF late on New Year's Eve, 1978, seeking to retrieve their lost test tubes for a new life outside academia. But they could not simply take the quickest technical route to their cloning goals. To avert legal complications, Genentech's plan was for Seeburg to make new growth hormone cDNA with mRNA extracted from human pituitaries supplied by Kabi and reclone it, and to speed the process by using his old UCSF clones as probes to screen the new clones. But the Kabi pituitary mRNA was badly degraded; when converted to cDNA it looked like just a smear on an electrophoresis gel, lacking any discrete bands. Cloning from the smear would not work.[28]

At this point accounts diverge about what actually occurred at Genentech. According to Goeddel and some others, Seeburg was depressed when he started at Genentech and working short hours. Around the beginning of February 1979, Heyneker asked Goeddel to help him. When Seeburg and Goeddel together were still unable to get large cDNA clones from the Kabi RNA, Goeddel says that they made new cDNA from good brain tumor RNA that Seeburg had brought with him from UCSF. With this RNA they produced and re-cloned the 550 base *Hae*III cDNA corresponding to the hormone from amino acid 24 to the end. This is how Genentech's *Nature* article reporting the accomplishment describes it. In 1980 the University signed an agreement with Genentech retroactively allowing the firm's use of the mRNA in this way.[29]

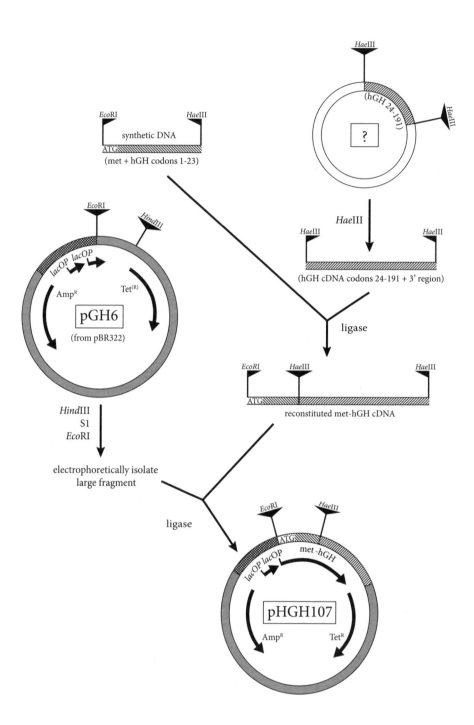

But according to Seeburg, after three failed cDNA cloning efforts from the Kabi pituitary smear, he and Goeddel decided to just splice the 550 base *Hae*III clone he had made at UCSF into the semisynthetic expression plasmid, which was now ready. This is the only major difference between the two accounts: whether the 550 base *Hae*III fragment of human growth hormone cDNA was isolated from the very same UCSF pituitary RNA *before*, or else *after* Seeburg came to Genentech. A great deal of money later depended on the distinction in a lawsuit between Genentech and the University of California. Perhaps the truth lies somewhere between these already subtly different stories. Either way, the material and knowledge imported directly from the publicly funded work of UCSF made up the greater part of Genentech's human growth hormone. As Kleid philosophically reflect after the 1999 $200 million settlement paid by Genentech to the University, the outcome was not entirely unjust, considering how much both Seeburg and the University had done to bring the protein to market. Except that of all the many drug firms that profited from that university work of 1977 and early 1978—and in 1999 about $1 billion in growth hormone was sold globally—only Genentech ever paid.[30]

And it was certainly Genentech's patent-protected method of building the first part of the gene with synthetic DNA that solved the expression problem holding back the UCSF group. As they had planned, the Genentech group joined the 550 base *Hae*III fragment of human growth hormone cDNA derived from the UCSF pituitary tumor to a short segment of synthetic DNA made by Itakura and Crea (the latter joined Genentech in late 1978), encoding the first 23 amino

---

Fig. 3.2. Genentech's semi-synthetic cloning procedure for human growth hormone in 1978–79. The *Hae*III cDNA fragment encoding amino acid 24 of mature human growth hormone and the 3′ remainder of the protein is joined by blunt end ligation to a synthetic DNA fragment encoding an initiation methione and amino acids 1–23; this in turn is inserted into a previously prepared expression plasmid containing a tandem *lac* operator-promoter (*lacOP*) and downstream DNA that terminates at an *Eco*RI restriction site positioned immediately before the native *lacZ* initiation codon. NB: *Hae*III digestion leaves blunt ends that can be joined to blunt ends left by S1 nuclease digestion of any other cut restriction site (in this case, *Hind*III), destroying both recognition sites in the process. The source of the *Hae*III growth hormone fragment, indicated by the question mark, was either a plasmid such as pHGH1 previously constructed by Peter Seeburg at UCSF or else one newly cloned from pituitary tumor cDNA by Seeburg and David Goeddel at Genentech, as disputed in court. (Not all enzymes and substrates depicted; see US Patents 4,342,832 and 4,601,980 and main text for further sources and details.) Abbreviations: Amp, ampicillin; met, methionine; hGH, human growth hormone; Tet, tetracycline.

acids of the 191-amino acid human growth hormone. In addition, at the beginning of the sequence encoding the first 23 amino acids, they added DNA coding for the amino acid methionine, which is necessary for the start of protein synthesis (but is often clipped off the protein afterward). This 77-base synthetic piece, including restriction enzyme sites at either end, was spliced to the natural cDNA fragment encoding amino acids 24 to 191 of the hormone and inserted into a plasmid downstream from another version of the high-output mutant *lac* promoter that Gilbert and Fuller had been trying to use for insulin (fig. 3.2). Here it was meant to be transcribed into RNA and translated into protein in place of the lactose-digesting genes. And it worked: when the *lac* operon was switched on, the bacteria produced significant quantities of human growth hormone, or hGH. (The initial, extra methionine was still attached, but the 192 amino acid version of the growth hormone seemed to behave just like the natural 191 amino acid version.) The Genentech team reported their triumph in a *Nature* article of October 1979. Seeburg's name appears as the final author in an interesting emulation of academic authorship practices; although Goeddel felt he did most of the work himself (as reflected by his first authorship) and Seeburg very little, Heyneker persuaded him to give Seeburg the honorific last authorship position—traditionally bestowed on the senior scientist whose resources enabled the work even if he did not "do" any of it.[31]

Meanwhile, without Seeburg, Goodman's UCSF group continued their effort to clone human growth hormone along the path already forged, now with Lilly funding. By June 1979 they had cloned the 800-base cDNA corresponding to the whole human growth hormone mRNA, using essentially the same approach Seeburg had reported in the December 1977 *Science* article on the rat gene (fig. 3.3). The 800-base insert from the *E. coli* transformed with the big human cDNA showed, from its sequence, that it included the code for the prehormone leader—the stretch of amino 26 acids cleaved off the hormone precursor in the pituitary to yield active hormone, including the methionine codon with which translation would in nature begin as well as some untranslated nucleotides preceding it. This oversized human pre-growth-hormone gene was cut out of pBR322 and spliced into another plasmid containing genes from *E.coli*'s tryptophan operon, specifically into a *Hind*III site near the beginning of the gene encoding one of the tryptophan-synthesizing enzymes (*trpD*). Now, when the operon was turned on by manipulating the growth medium, the bacteria produced large quantities of a protein consisting of the first part of that bacterial enzyme fused to all of the pre-growth-hormone. Since neither the prehormone leader nor the bacterial *trpD* part of the fusion protein could be clipped neatly off, this genetically engineered microbe could not produce medically usable hormone. Still, the exercise could be seen as another step toward producing mature human growth hormone in bacteria; presumably, the UCSF group intended to shorten the hormone-coding portion and line up the first codon with an initiation codon by a multistep, exonuclease-based procedure like that on which Gilbert's insulin cloning strategy had been based. An article describing this UCSF work, important from a scientific perspective in describing sequences controlling the translation of human growth hormone in nature, but very unfinished from the perspective of pharmaceutical production, appeared in *Science* magazine in August 1979.[32]

Shaky Steps From Bench to Bedside

Both the UCSF *Science* article and the Genentech *Nature* article of 1979 have been astutely described by business historian Maureen McKelvey as publicity exercises, propaganda offensives in the now-fierce contest between the University team and Genentech. As Genentech lawyer Thomas Kiley later recalled, "there was a time when you couldn't get a publication out of Genentech that didn't have little bottles all over it and indented into it," metaphorically speaking, "to make the point we were moving from the bench to the factory." As its boastful, plain-language conclusion illustrates, the Genentech *Nature* article was surely one of those intended for business audiences as well as the scientists from whom the young Genentech scientists sought credit:

> Using a novel combination of chemically synthesised DNA and cDNA, a recombinant *E. coli* strain has been constructed which produces hGH in large amounts. This is the first time that a human polypeptide has been directly expressed in *E. coli* in a non-precursor form.[33]

Thus growth hormone surpassed the earlier somatostatin success because the gene for that hormone was expressed directly as a fusion product and later freed. The competing growth hormone articles were preceded by dueling press conferences on July 11, 1979, a few days after Genentech had submitted both its patent application (covering semisynthetic expression vectors generally, not just growth hormone) and its report on the work to *Nature*—and a month before its acceptance, which according to scientific etiquette should be the starting point for releasing results to the public. Catching wind of the Genentech's impending announcement at the eleventh hour, UCSF hastily cobbled one together for the same day that described their own forthcoming paper, by this time peer-reviewed and accepted. Both events represent examples of the then-controversial

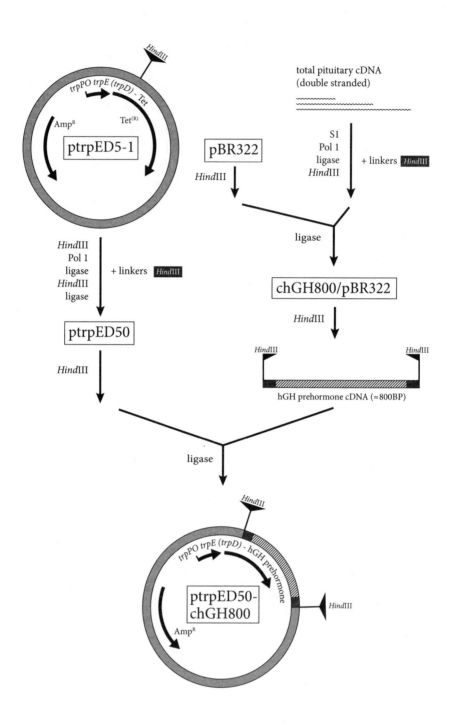

practice of "cloning by press conference," or release and reporting of scientific results before their presentation and full critique in peer-reviewed formats, mainly for public relations and financial markets impact.[34] Effectively Genentech was setting the tone that academic competitors had to emulate, to play in the same commercial register for audiences beyond science. Academic entrepreneurs like Rutter and Gilbert readily picked up the new tune.

In early 1980 Genentech began preparing for its October initial public offering (IPO) of stock on the strength of its cloning successes with insulin, growth hormone, and especially interferon (chap. 4). In exchange for $350,000 and modest future royalties, the university granted Genentech free use of any tangible materials made by the postdocs, such as the pituitary mRNA—but not patents arising from their work, which had already effectively been sold to Lilly. As noted this agreement removed the threat of litigation by UCSF over Genentech's use of the UCSF pituitary mRNA as a source of cDNA for recloning and brought money to the university. Relationships between the UCSF and the Genentech researchers warmed until the next round of litigation—this time involving Lilly. For the moment peace prevailed. Around this point Swanson must have started thinking seriously about marketing growth hormone as a drug in competition with the NPA; as noted he renegotiated the Kabi contract so as to give the North American market exclusively to Genentech in exchange for lower royalties elsewhere.[35]

The early collaboration with Kabi helped Genentech learn the pharmaceutical ropes, particularly in scale-up and manufacturing. Already in September 1978, when the growth hormone plan had just been sketched, Kabi sent an antibiotic fermentation expert to San Francisco, and Kabi personnel were present in

---

Fig. 3.3. The full-length human growth hormone gene cloned and expressed as a fusion protein in 1979 by Joseph Martial and others at UCSF, after Seeburg's departure. A double stranded cDNA of about 800 bases in length, corresponding to essentially the entire human growth hormone messenger RNA and encoding the mature hormone as well as a 24-amino acid prehormone leader as well as 3′ untranslated regions, is rendered blunt-ended by enzymatic treatment and joined to synthetic *Hind*III linkers. This cDNA is inserted into the *Hind*III site of an expression plasmid containing the 5′ region of the *trp* operon, previously prepared by addition of an extra linker to shift reading frame, early in the coding sequence of the *trpD* gene. In the final product a trpD-pre-growth hormone fusion protein is translated from *trp* transcripts. (Not all enzymes and substrates depicted; see *Science* 1979;205:602–7 and main text for further sources and details.) Abbreviations: Amp, ampicillin; hGH, human growth hormone; Pol 1, bacterial DNA polymerase 1; S1, nuclease S1; Tet, tetracycline.

San Francisco for much of the next two years. During 1979 the Genentech-Kabi manufacturing group in San Francisco grew small batches of engineered bacteria, froze them, and flew them back to Sweden for work on preparing the pure growth hormone from the cultures. As they had with insulin, the young biologists at Genentech re-engineered *E. coli* to produce growth hormone (still with the extra methionine, to be brand-named Protropin) from the high-output *trp* promoter, which doubled yields but left much of the recombinant hormone in insoluble form. Genentech developed their own method of recovering hormone from the refractile bodies, while Kabi continued to extract only the soluble portion of the hormone. In mid-1980 Kabi moved its recombinant growth hormone fermentation and production research to Sweden, marking the divergence of the two firms' production methods and Genentech's independence.[36]

While Genentech's molecular geneticists worked hard in the background to engineer a new expression system, so that the bacteria would process off the methionine and secrete "true" 191-amino acid growth hormone, Swanson pushed to get the 192-amino acid Protropin approved by FDA and on the market. In early 1981 the first human test of the product highlighted the shortcomings of the academically trained biologists at Genentech, who had renatured and purified growth hormone from the refractile bodies until they could detect no other protein in their gels. Genentech employees served as guinea pigs in this Phase I safety study (in FDA language), and to their surprise and dismay, those receiving their own product developed inflammations while those given the brain-derived Kabi product did not. The problem was eventually traced to contaminating bacterial membranes; one can easily picture why, for molecular biologists (even protein chemists) steeped in the field's DNA-RNA-protein paradigm, the possibility of stray lipids (fats) did not register on the intellectual or technical radar.[37] And it was not just the firm's molecular biologists that were naive about the business of making pharmaceuticals. In 1982 Swanson threatened to sue Kabi when the Swedish firm planned to notify the FDA that some rats tested with their shared growth hormone preparation, about to enter full clinical trials, had developed abnormal growths.[38] Covering up such adverse drug information could easily have ruined Genentech, and if patients were eventually harmed might have led to Swanson's criminal prosecution.

Funding clinical trials both for officially diagnosed pituitary dwarfism, and also for juvenile shortness without abnormally low growth hormone levels, Genentech had gathered FDA-required evidence on the drug's efficacy and safety by late 1984. In some of these clinical studies, the recombinant Genentech product appeared more likely to raise allergic reactions, even though purification procedures were now much improved. Understandably, the FDA did not move to approve the recombinant hormone drug; after all, it offered no medical benefit over the existing brain-derived drug, the supply of which was adequate for bone fide cases of pituitary dwarfism, while it presented extra risks of allergic reaction. But fortune swung in Genentech's direction in early 1985, when a few cases of the deadly condition Creutzfeldt-Jakob disease (of mad cow notoriety) were linked to brain-derived growth hormone. In October 1985 the FDA approved recombinant growth hormone for pituitary dwarfism because allergic reactions now represented the lesser risk.[39]

In its first full year, 1986, Protropin sales reached over $40 million—about double what they would have been if only the officially diagnosed patients previously receiving pituitary hormone had used it. By October 1986 there were already press reports that the drug was being used nonmedically by cheating athletes, and possibly being overprescribed for healthy children whose parents sought to give them a height advantage. Publicly tarred by association with ethically dubious "cosmetic endocrinology," Genentech had truly arrived as a pharmaceutical firm. In 1987 Lilly paid it the ultimate compliment of imitation by launching its own human growth hormone—Humatrope, 191 amino acids and lacking the extra methionine—prompting Genentech to sue Lilly for infringement (allegedly using the semisynthetic vectors licensed to the firm only for insulin), and the University of California to sue Genentech. This multiparty lawsuit, involving insulin as well as growth hormone, was complex and lengthy. In 1995 Lilly settled by paying Genentech $145 million, and as mentioned Genentech settled the university's claims to growth hormone with a $200 million payment in 1999.[40]

Despite the unexpected competition from Lilly Protropin sales grew strongly, thanks to the marketing expertise of Kirk Raab, hired in 1985 from Abbott pharmaceuticals as Genentech's Chief Operating Office. In 1999 Genentech paid one of the largest fines to date for illegal marketing, for example by running "screening" programs in primary schools (both directly and by funding the nonprofit Human Growth Foundation patient group) that sent "information" on growth hormone to parents of children in the 5th height percentile. There was also ample evidence that off-label prescribing for unapproved uses accounted for a large proportion of this product's sales. The success of Genentech's first product already highlights one of the difficulties in assessing the social value of the early recombinant drugs. When much of a product's sales are accounted for by medical uses other than approved indications—that is, uses for which there does not exist solid evidence from controlled trials that clinical benefits out-

weigh harms—sales figures cannot be taken to reflect the benefits of the innovation. As we will see, this problem was not limited to human growth hormone:
illicit marketing and significant off-label sales also occurred with recombinant
interferon alpha (chap. 4) and, with dire results, erythropoietin (chap. 5).[41]

Reflections: The Scientific Ethos
and the Biotechnology "Revolution"

As we noted at the outset, there has been much talk of revolutionary changes
brought upon academic biology and the drug industry. Indeed, a new way of
scientific life was taking shape in the labs of Genentech. Since it would set the
pattern for many of the early biotech firms that young biologists were joining in
the early 1980s, it would be appropriate here to discuss. One claim that has been
made is that the biotechnology industry's rise owes much to the political radicalism of the late 1960s and early 1970s, as young biologists rejected the socially
irrelevant basic research of academic life science to pursue life-saving advances
within the drug industry. This view seems a stretch, given that, to leftist sensi-
bilities in the 1970s, pharmaceuticals were a morally dubious enterprise bearing
no resemblance to risking one's life to end the killing in Vietnam or racism in
the American South. FDA Commissioner Goyan was only echoing mainstream
perceptions when he blamed big pharmaceutical firms for a national "overmedication" problem, particularly with tranquilizers and antibiotics, in national
media. Likewise, Senator Edward Kennedy was pursuing popularity points
with his aggressive mid-1970s hearings "to restructure and reform" the industry. One of Kennedy's main complaints was that the drug industry devoted most
of its research efforts to "me-too drugs" or redundant medicines, copycats barely
different from established products, rather than genuinely novel contributions
to health care. (This is a critique worth remembering in the context of recombinant growth hormone, developed by Kabi and Genentech when experts considered supplies of pituitary growth hormone sufficient to meet existing need.)[42]

Furthermore, fitting with leftist antipathy to industry generally, the life scientists who most actively opposed the Vietnam War—like the Science for the
People group—mostly opposed genetic engineering. And there is no evidence
of special political engagement among the young scientists who joined Genentech and thus helped set the tone for the early biotechnology industry. Like those
joining other firms in the early days, many of them report initially thinking little
and knowing less about the drug industry. So the first biotechnologists were no
Hippies, Yippies, or Freedom Riders—or even idealist reformers from within.

Rather, these young scientists had high career ambitions and were very competitive. Indeed they were in too much of a hurry to get ahead scientifically, in the eyes of some professors. As Rutter has said of two who joined Genentech early:

> Axel Ullrich . . . was very ambitious and sophisticated, driven to success. In my experience, he and Seeburg were the first to be concerned about accreditation and attribution. Previously, people were more trusting and less concerned with who got credit, at least before the experiments were done. They were more agreeable to cooperative studies. These individuals [Ullrich and Seeburg] really foresaw the importance of these studies, and the notoriety that would come from carrying out experiments like this. They essentially prenegotiated their roles in the experiments that were to be performed and therefore prenegotiated essentially their positions in the authorship on papers.[43]

High scientific ambitions characterized the others too, like Herb Heyneker, who returned to Genentech from his job at the University of Leiden because the Netherlands imposed an informal moratorium banning the cloning experiments he wanted to do, and thus guaranteed he would not do them first. Of course Heyneker helped recruit Kleid, whom he knew from academic circles, and Kleid recruited Goeddel. Compared with Seeburg and Ullrich, these three had fewer initial doubts about working in an industrial context, but they were just as competitive, and they shared with the others the view that cloning advances, whether the of *lac* operator (Heyneker's early project with Boyer at UCSF) or of a protein hormone (Seeburg's goal even before he arrived at UCSF), were themselves fundamental contributions to life science that would win fame or "notoriety." And in this the postdocs who went on to do Genentech's science did not differ from the California professors they worked with.[44] The chief difference between the younger generation and many of the older was about how the credit should be distributed.

Still, there may be a germ of insight in relating the radical break of these young Turks to youthful political rebellion in the late 1960s and 1970s—not directly, but through a common underlying cause. In his analysis of the youth revolt that swept much of Western Europe in 1968, sociologist Pierre Bourdieu drew special attention to the situation of some of the revolt's main leaders, students at the elite universities. The postwar expansion of the universities had made state-funded academic careers a safe path to the upper middle classes. But when universities stopped growing at the same pace, students found that there were far too many of them for the jobs they sought, leaving them feeling exploited by the professors and the educational system.[45] As we have seen, by the early 1970s the seemingly endless expansion of funding from NIH and other federal agencies—which had made American biomedical science the envy of the Cold War world—had stalled. Grant success rates were declining, and senior scientists with a track record from large grants were more likely to receive renewed grants than any junior scientist was to win his or her first. In a situation of steady-state academic hiring, very few of the half-dozen postdoctoral fellows typically passing through an important senior scientist's lab group in a given year could expect to win a full academic position. So the postdocs rebelled against the professors and sought to set up shop on their own.

In the early days of Genentech, where Swanson needed them so badly, under Boyer's sympathetic oversight they built a better place for themselves to do science, something like a postdoc's republic where one of them even sat on the executive board (rotating the place in egalitarian fashion with the other scientists). Culturally as well as economically, these young biologists carved out a new domain adjacent to universities, suited to a young person's restless temperament and cracking pace of work and relative lack of other responsibilities—and also suited to their particularly high scientific ambitions.[46] Thus, apart from disregarding the old disdain for engagement with politics and practical application (which had become unfashionable among university scientists generally in the 1970s), I would argue that these young biotechnologists were idealists in remaining true believers in the scientific culture that had grown up around the Cold War ideology of basic research.

Established professors and junior scientists of course share the values and codes of behavior governing the scientific fields to which they both belong. But they differ in their interpretation of certain rules. Common throughout academic science then (and in some places now) was something like the code described by sociologist Robert Merton, an architect of the Cold War research ideology (chap. 1). According to this code, good scientific behavior requires a high degree of impartiality in various ways, and also "communalism"—a cooperative and egalitarian sharing of information and materials with peers that may clash with immediate self-interest. Impartiality and communalism were obviously under strain in the events described here. But the most fundamental rule, traced by Merton to the seventeenth-century founders of the British Royal Society, is that scientific credit goes to those who first fully describe their experiments and other work, so that other scientists can verify and build upon it. Nothing was (and, I think, still is) more important to academic scientists than the credit won from authorship of scientific papers. In the 1970s some of the biologists who, thanks to more than two decades of NIH largesse, had come to manage big laboratories with large budgets (like Bill Rutter's 30-strong lab group), were accustomed to apportioning work and resources among individuals. They sometimes exercised this control undemocratically to maximise the credit garnered by the group bearing the professor's name, even at the expense of particular juniors.[47] So the very same actions justified by a lab chief's appeal to communalist ideals, such improving the group's collective chances against a competitor, could be critiqued as self-serving and exploitative from the perspective of a junior scientist of particularly high ability and ambition. These generational frictions of crowded laboratory groups, especially in lean times, seem to have grown acute at UCSF in the mid-1970s.

Ullrich and Seeburg, for example, resented being assigned by Goodman to projects they regarded (respectively) as "esoteric" or long shots, because they harbored ambitions to be the first to clone a human hormone. Esoteric research projects are those that, even if they succeed in their own terms, will never win great accolades. Long shots on the other hand can greatly reward both postdocs and lab chiefs, but in the most likely event of failure, the junior pays the higher price: forfeiting his career by achieving nothing at the crucial postdoc stage. And once an achievement was ready to publish, senior professors used their power of the pursestrings to influence authorship of the papers issuing from the lab, sometimes paternalistically to aid their juniors' careers, but sometimes less benignly. Many a professor regarded it as their right to be included as an author (typically, the final author) on any paper where the work was done in his lab, or under his grants, regardless of personal contribution to the actual lab work or writing. Postdocs and graduate students did not have the power to refuse such expectations, and if they lacked what Rutter called "trust" in the professors, this could be galling. An egregious example was, for Seeburg, Goodman's efforts to take credit for the growth hormone cloning, which had occurred despite his discouragement and mostly during his sabbatical absence.[48]

Thus we might understand Genentech's policy of allowing open publication by its scientists not just as a tactic to advertise the quality of the firm's science in leading journals, but also as a product of postdoctoral scientists' sense of justice and overriding drive to build a reputation quickly. As Ullrich recalled, early recruits like Seeburg and himself wanted to publish, partly because they needed the credit in order to keep open an avenue for return to academia. In order to retain them, Boyer persuaded Swanson to let them build "a university-like atmosphere" with opportunity to pursue curiosity-driven projects. And this they did, succeeding not only in making Genentech a respected source of what could be classed as basic research, such as Ullrich's work on insulin receptors and Goeddel's on the growth-regulating oncogenes, but in building academic career options for young scientists in the firm. Seeburg and Ullrich, for example, both eventually became heads of prestigious, publicly funded research institutes. Even for the less academically oriented early recruits like Goeddel, competition for scientific credit via prominent publications was a prime driver, so much so that cloning a protein that was also a promising drug excited him largely because medical promise meant publication in a better journal. Similarly, Genentech scientists shared cells and materials from published work like the better academic labs, eventually dedicating several staff to handling materials requests, and welcomed academic scientists on sabbatical, despite the recognized risks of disseminating proprietary techniques. Patent applications were not permitted to retard submission to journals, often leaving legal staff only a few days to write and submit them.[49] Even allowing for the rose-colored glasses with which many early Genentech employees regard the first days there, one can read (a little speculatively) into their corporate culture an ideal scientific community more perfectly reflecting the core values of basic molecular biology, in reaction against an actual university atmosphere distorted by professorial hypocrisies. In this antiauthoritarian scientific communalism, where everyone was an Indian and no Chiefs were welcome, the rise of biotech reflects something of the spirit of 1968.

What of the opposite critique that, diverted by the temptations of riches, biologists in academia (and in the biotechs) turned their attention away from the fundamental and ultimately more valuable research that had previously occupied them, to merely practical drug projects promising quick returns? Here and in other chapters I point out junctures at which the behavior of these scientists was driven by the quest for scientific credit, but there are other grounds to contest this view without debating the weight of different motivations in the past.[50] For a basic assumption of this critique is false: the idea that cloning hormone genes to express in bacteria was a *merely* practical goal diverting molecular geneticists from the type of fundamental research on which their field had previously focussed. Indeed, the assumption is wrong on two counts. First, in the 1970s, cloning a gene for a human protein so that it would be expressed in bacteria was an important frontier in what the field already regarded as basic research. Simply developing techniques to enable biologists to pick out a particular expressed sequence, like a needle from a haystack of RNA, and to clone and sequence it, was news worthy of *Science* and *Nature*. Likewise sensational as a basic advance was showing that synthetic genes functioned in bacteria, no less so when Genentech made insulin this way than when the Boyer and City of Hope groups did the same thing with the commercially worthless *lac* opera-tor.[51] Furthermore, beyond advancing technique, as noted earlier cloned cDNA sequences allowed one to explore theoretical problems of how genes evolved, and how higher organisms regulated gene expression at the level of translation—and potentially also transcription (if cDNA was used to retrieve neighboring chomo-somal sequences, as indeed the UCSF group did). So, at the time, cloning and expressing hormones in bacteria was cutting edge life science—not something less, or different, just because the protein might have commercial value. Sci-entists did not need to be diverted by the promise of riches, because the same work would bring them traditional scientific credit. After all Seeburg had been working on cloning growth hormone for two years, before he or his professors or Swanson and Boyer thought about selling it. (Later, as I will argue in chap. 4, certain kinds of money began actually to *confer* scientific credit.)

But the view that commercial reward corrupted and diverted basic research in life science rests on a second, deeper false assumption: it takes for granted the notion that the "basic research" that molecular biologists were doing, even before they considered cloning hormones, was "pure" in the sense that it was far divorced from application. As we have seen (chap. 1), apart from the Cold War political utility of science that could be advertised as a free quest for natural truth, and pure or fundamental at least in the sense of lacking direct military application, there was another kind of utility supposed of NIH-supported life science—at least by the Congressmen voting public funds for it. As an explicit rationale for this agency's huge budget expansion during the 1950s and 1960s, the laboratory science funded by NIH was always already "preclinical" in a literal and strong sense: just a step removed from clinical utility. That small step was to be left to private enterprise. And as leftist critics like Science for the People liked to point out, NIH funded far more research into the cellular and genetic mechanisms of cancer, to enable new clinical interventions, than into the imme-diate and indisputably more cost-effective strategy of cancer prevention through reduced exposure to pollutants.[52] Even if scientists sometimes felt that the medi-cal promises they wrote into NIH grant applications were tall tales, their funders were wiser than they realised. After all, when Paul Berg switched from study-ing gene regulation in bacteriophage to gene regulation in monkey viruses he was not just studying DNA-protein interactions as an abstract problem.[53] He was following the NIH money and actually working on cancer, almost incidentally inventing a way to move cancer genes in and out of bacteria (and monkey cells) that also enabled commercial genetic engineering.

Thus genetic engineering to make potentially therapeutic proteins began without concrete interest from the drug industry or other private money on the table; it followed as if naturally from the preclinical biology research NIH had been funding since the early 1950s. In 1984 the Congressional Office of Technology Assessment acknowledged this as obvious.[54] That so many biologists began looking for commercial sponsorship in the late 1970s may have more to do with the NIH funding plateau, and more generally with the end of the early Cold War's science funding regime, than deliberate policy. Certainly some entrepreneurial biologists, like Rutter and Gilbert and Boyer, began to perceive the change in society's reward structure; even the National Cancer Institute's still-strong budget could not be expected to fund everything. So they moved their research "downstream" from strictly exploratory work on gene expression toward saleable proteins before 1980. Some, like Rutter, tried extending themselves to commercial goals in a way that would preserve the hierarchical structures of the academic lab groups they operated. David Goeddel criticized this aspect of Genentech's arch-rival Biogen (Gilbert's company), contrasting that firm's professorial emphasis in its early days to Genentech's more cooperative egalitarianism, where he and the other "young scientists were the center of things." But young or old, these responses answered the same problem: how to stay at the forefront in molecular biology with limited public funding. Celebrated policy landmarks like the 1980 Bayh-Dole Act, encouraging academia to expand into the domain previously reserved for the drug industry, underscored the writing that many scientists already saw upon the wall.[55]

CHAPTER FOUR

The Interferon Derby

Markets in Credit, Tournaments of Value

Insulin may have been the most famous protein in 1978, when Genentech claimed the laurels in the race to clone it, but the one most on the public's lips that year was *interferon*. Discovered 20 years earlier as a mysterious hormone-like substance released by animal cells when attacked by viruses, interferon had already passed through one cycle of hope and disappointment as a magic bullet against viral infections. The virus theory of cancer, so influential in the late 1960s and 1970s, had led biologists to test it on cancer in mice and then in a handful of human patients, with promising outcomes. These hopeful but inconclusive results caught the attention of American cancer research lobbyists and philanthropists in the middle 1970s, stimulating further research. But it wasn't until 1978 that interferon rose to global celebrity status when major newspapers and magazines from *Life* to *Der Stern* picked up on the American Cancer Society's unprecedented decision to purchase $2 million of the stuff for more systematic clinical trials.[1]

Interferon offered hope when it was politically required: the National Cancer Institute and the "cancer establishment" it funded was under fire after seven years and $5 billion worth of America's War on Cancer—to skeptics an unwinnable "medical Vietnam," as Donald Kennedy so aptly put it. Suddenly this "natural body substance" (implying much milder side-effects than cytotoxic, cell-killing cancer chemotherapy) became an NIH flagship project, too. Suddenly letters from patients begging for the miracle drug flooded hospitals and labs associated with interferon research. None was available for purchase, but molecular biologists were promising very publicly to make it available through genetic engineering.[2] So in 1978 the long anticipated cure for cancer, in the shape of interferon, was what the public most wanted from the new biotechnology. And the entrepreneurs of biotechnology were eager to sell it—and with it their firms, and their whole industry—to the public. As I shall argue here, the mutually re-inforcing coincidence of the interferon mania with the advent of recombinant DNA technique played a major role in generating the first wave of biotechnology firms. If there was ever a gold rush in genetic engineering, it was among the biologist-entrepreneurs who pursued interferon's commercialization, skillfully solving a scientific problem while cultivating public expectations and reshaping the investment markets that paid for it.

When the interferon cloning race among molecular geneticists began, inter-feron research was already evolving from a cottage industry, governed by a tra-ditional gift economy, to a fast-paced field grown more technology-intensive and competitive through recent publicity and funding. The interferon protein, first named and characterised phenomenologically by its interfering effect on virus infection, was no longer dismissible as "imaginon" or "misinterpreton." As bio-chemists came to grips with it, interferon revealed itself to have a multiple iden-tity. Now there was leukocyte interferon (the main interferon derived from white blood cells), fibroblast interferon (from connective tissue cells in culture), and rarest of all, immune interferon (from particular white blood cells)—each dis-tinguishable physically as well as biologically. Soon these would be renamed alpha, beta, and gamma interferon. And new, more sensitive biochemical meth-ods were starting to yield amino acid sequences for these proteins, suggest-ing that leukocyte interferon itself comprised multiple similar molecules. Since interferon effects appeared strongest in the species of origin, like many protein hormones, only human interferon would do as a drug—and it could be made by cultivating human cells on a large scale.[3] Thus academic biomedical science had already learned much about interferon, and was already moving toward manu-facturing the precious substance(s) when the cloners and their investors came on the scene.

Naturally, private investors in biotech were drawn to interferon by the level of popular enthusiasm and the enormous value attributed to it by some media coverage (up to $50 million per gram, or $250,000 for a course of treatment). Multiplied by the hundreds of thousands of cancer patients treated each year in the US alone, these predictions in the media implied astronomical market potential—certainly in the billions.[4] And of course, interferon's potential was not lost on the major pharmaceutical firms, some of which had already invested in interferon research and could be counted on to pay handsomely for any way to bring the miracle cure quickly to market. Thus in 1978 human interferon was the genetic engineer's top prize: exceedingly rare, exceedingly valuable, sure to win fame, and sure to attract the toughest competition. In this chapter we focus on the effort to clone leukocyte or alpha interferon—the interferon with the most promise as a cancer drug at the time, the first made in bacteria, and the first genetically engineered interferon drug marketed. Fibroblast (beta) interferon and immune (gamma) interferon would be cloned almost simultaneously, in a contest involving the same competitors with the same set of techniques, although different winners.

Front Runner

At Genentech, fiercely competitive David Goeddel set his sights on this "magical substance" that was the "ultimate project" for a cloner. But he needed a source of messenger RNA to make cDNA, and Genentech needed money to pay for the project. At roughly the same time (mid-1978) that Genentech announced it had expressed human insulin and formed partnerships with both Lilly and Kabi, Genentech CEO Swanson began seeking a similar contract that would pay the firm for cloning interferon in installments, and eventually royalties. Although the final contract was not signed until January 1980, Swanson soon resolved both the finance and mRNA source problems through an agreement with the Swiss multinational Hoffmann–LaRoche (Roche). Under its terms, Genentech would collaborate with biochemist Sidney Pestka at Roche's New Jersey research labs. Pestka had long been studying interferon and was starting to obtain (but not publish) some amino acid sequence from leukocyte interferon.[5] The collaboration was limited: Pestka would independently try to clone the interferon himself, but would also supply Genentech with the same mRNA he was using, which he sourced from a cultured human cell type called KG-1. Pestka had acquired the KG-1 cell line from an NIH researcher, Robert Gallo, who discovered that it overproduced interferon. Gallo had in turn acquired the cell line from his long-time friend and colleague David Golde, a UCLA hematologist and oncologist who had collected it from a 59-year-old leukemia patient. (Golde was known in scientific circles for a remarkable collection of human cancer cell lines that produced rare growth factors in culture; he would soon become world-famous because of the lawsuit brought by another one of his cancer patients, John Moore, over ownership of cells taken from him.) In December 1978 Goeddel commenced a solid year of what Genentech attorney Tom Kiley recalled as "100-hour weeks" trying to clone interferon alpha, diverted only briefly in early 1979 by helping Peter Seeburg clone growth hormone (chap 3).[6]

For most of 1979 Goeddel found little but frustration. Pestka sent him cell extracts or RNA from New Jersey, which Goeddel was supposed to turn into cDNA, clone, screen, and then consult Pestka for confirmation that cloned sequences were interferon. With Pestka's material he applied standard procedures to separate mRNA from other RNA, and used a sucrose density gradient to enrich the sample for mRNA of the approximate size. Then he treated it all with reverse transcriptase, ran the cDNA on a gel, cut out a section corresponding roughly to 600–1000 bases as expected for messages encoding a protein of interferon's size, and recovered the cDNA from the gel piece. Then he followed the Maniatis and Eftstratiadis procedure of G-C tailing and insertion into pBR322's ampicillin resistance site. But how to screen all the ampicillin sensitive, tetracycline resistant bacteria carrying a piece of cDNA—potentially from almost any gene expressed by KG-1—to identify the few encoding interferon?

Pestka was using a biological assay to help him find interferon clones (see below), but Goeddel depended strictly on the molecular screening method of differential hybridization. Here every clone is tested with a heterogeneous probe that hybridizes with the target gene and another probe that, ideally, is identical except that it lacks sequences from the target gene; for example, cDNA made from KG-1 cells before virus challenge, when they make little interferon, and after virus challenge, when they are "induced" to make the protein in quantity. Through comparing the two outcomes, this procedure identifies cloned sequences expressed only (or mainly) in induced cells. Predicting that even in induced KG-1 the interferon message would be rare, perhaps rarer than 1 in 1,000, Goeddel expected to look at many thousands of clones before he found one coding for interferon. "Brute force" is the apt description molecular biologists give this type of highly repetitive mass screening procedure.[7] As colleague Dennis Kleid recalled those 1979 experiments:

> Dave would make cDNA libraries from the leukocyte cells and end up with hundreds of thousands of colonies which we would then screen to see if we could find one that had the interferon messenger RNA. All of us would poke these colonies for him. You can't imagine how long it takes to take a stack of forty petri plates, each with hundreds and hundreds of colonies on it, poke each one with a toothpick, and make an array so that you could screen them.[8]

After looking at perhaps ten thousand colonies in this way, Goeddel would decide the "library" or collection of transformed bacteria from one batch of cDNA was no good and start again with more KG-1 material. But after a few cycles like this—and after talking with some of the younger scientists at Roche—Goeddel began to think Pestka was supplying him with "lousy RNA," too degraded to clone. He complained to Roche management, after which his cloning results seemed to him much improved.[9]

Meanwhile, back in New Jersey, Petska's group was also cloning from the KG-1 mRNA by a nearly identical method, using G-C tailing and inserting the tailed cDNA into pBR322's *Pst1* site. But Petska's screening method was different. First he employed a cruder molecular screen, probing arrays of transformant colonies with radioactively labelled RNA from induced cells mixed with a great excess of unlabelled RNA from noninduced cells (so that genes expressed specifically in induced cells would hybridize most strongly with radioactive probe). Ten percent of Petska's transformants showed some degree of signal, that is hybridization, indicating a sequence from genes expressed preferentially in induced KG-1 cells. These Pestka screened further using a biological assay for the specific presence of interferon messenger RNA. It had recently been found among his fellow interferonologists, who had always depended on biological assays, that interferon message could be detected by injecting extracted RNA into frog eggs in tiny dishes of water, then collecting that water the next day and testing it for interferon protein as measured in a standard virus-protection bioassay with cultured cells. Pestka used this to test clones for interferon sequence by immobilizing plasmids carrying DNA inserts on filter paper, allowing mRNA from interferon-producing KG-1 cells to hybridize, and then washing off the mRNA that stuck for the double bioassay. If a plasmid retained mRNA that made frog eggs produce interferon, it presumably carried interferon gene sequences.

This procedure was labor-intensive, so Pestka tested ten transformants at a time, pooling their plasmid DNA on the filters. His twelfth pool yielded consistently positive results, so he subdivided the pool and repeated the bioassays to find the clone in it that hybridized with interferon message. It actually contained two positive clones, but both carried cDNA too short to encode a full interferon gene. He sequenced the longer one and found that it agreed with his own unpublished amino acid sequence, showing it to be a piece of the interferon gene. At this point, perhaps around the time that Roche finally signed a contract with Genentech in early January 1980, he began to cooperate more closely with Goeddel.[10]

Somehow, in the mean time, a partial amino acid sequence from the interferon protein had fallen into Goeddel's hands and become the basis of his molecular screening method (Heyneker recalls scribbling a sequence from a conference slide by a Dupont researcher; in any case it was later confirmed by Pestka). At the time there was no straightforward way to clone a gene with only the protein product in hand, without access to the nucleic acid sequence (at least for proteins too large for complete synthesis of encoding DNA, unlike tiny somatostatin or insulin, and not abundant enough to enable the production of antibodies).[11] The problem with using amino acid sequence information to predict the interferon gene's sequence, and perhaps to make a matching synthetic DNA probe, is that there is no one-to-one relationship between the 64 possible nucleic acid triplets or codons and the 20 amino acids making up proteins. But Goeddel found two short stretches of the leucocyte interferon protein sequence consisting of low-degeneracy amino acids, and from these created two small sets of synthetic DNA oligomers 11 bases long ("11-mers") covering the possible sequences of these two parts of the interferon gene. These he mixed with mRNA from either induced KG-1 cells or noninduced cells, and added radioactive nucleotides and reverse transcriptase. Under these conditions, the single stranded synthetic DNA would be expected to bind to matching sequences in the mRNA—particularly to interferon message (but also to sequences from other genes that happened to match)—and these 11-nucleotide double stranded pieces would then be extended by the transcriptase to make longer radioactively labelled DNA probes from the RNA template.

Of 500 KG-1 cDNA transformants screened by this innovative method, Goeddel found 30 possible interferon clones. At this point in early 1980 Petska gave Goeddel his short interferon clone, which Goeddel used to probe for matching sequence in another 400 transformants, yielding 9 more possible interferon clones. Among these 39 he found one containing a 1000-base cDNA insert, long enough to represent the whole interferon message. The Genentech group sequenced it and found that it fit with the amino acid sequence of Petska's leukocyte interferon. Goeddel now had his confirmed clone. But he was not alone at the finish line.[12]

Another Dark Horse

In the second half of 1979 and early 1980, Goeddel knew he was competing with similarly dedicated and youthful cloners in multiple commercial interferon efforts, as well as others in academic labs seeking only scientific glory. He must have known that these competitors included a team at Cetus, Genentech's San Francisco area rival and only real predecessor as a small biotech firm (founded in 1971 by physician-entrepreneur Peter Farley, biologist-businessman Ronald Cape, and Berkeley physics Nobelist Donald Glaser). In 1979 Cetus had developed a way of extracting interferon mRNA by fusing certain human cells with semicancerous cultured mouse cells, and was cloning human (hopefully) cDNAs from such hybrid sources into pBR322 by essentially the same G-C tailing method as Goeddel. They were also using a similar molecular screen in tandem with a screen for transformed bacteria capable of protecting cultured cells from virus infection—a direct bioassay for bacterial interferon production that was fraught with problems from bacterial toxins, as Biogen's simultaneous efforts would find. They did isolate, sequence, and clone both leukocyte and fibroblast interferon genes, and managed to engineer hamster cell lines to produce interferon, eventually winning valuable patents and bringing bacteria-made fibroblast interferon (interferon beta) to market in 1993. But Cetus, spread thin with projects related to energy and chemical production as well as pharmaceuticals, was too slow with leukocyte interferon.[13]

In mid-January 1980—luckily for Genentech one week after Roche's final commitment to the biotech firm—a Boston press conference announced the cloning of leukocyte interferon by Genentech and Cetus's mutual nemesis: Biogen. It triggered a media storm, thanks, according to the announcing scientist, to advance publicity by Genentech and Cetus: "the stage had been set largely by West Coast entrepreneurs, and when the curtain rose—albeit on the wrong cast—the success of the event was guaranteed." On stage in command of the surprise event was Wally Gilbert himself, Biogen's most prominent scientist.[14]

Charles Weissmann, the Swiss biology professor standing beside Gilbert, had originally entered the interferon cloning race strictly for the scientific honors. In mid-1977, while Boyer and Swanson's small Genentech crew were struggling to finish cloning somatostatin, and the Gilbert and UCSF groups were competing to clone insulin-coding cDNA, Weissmann happened to have a long chat with Yale biologist Peter Lengyel, an interferonologist and old friend from their shared postdoctoral days in Severo Ochoa's New York laboratory. Lengyel told him in detail about the state of play in interferon research, including the sensitive new frog egg assay. The two developed a plan to clone mouse interferon. It was no great departure for Weissmann, whose work at the time focused on the basic mechanisms of gene expression in eukaryotes, and who in mid-1977 had just made important contributions to the discovery of split genes by cloning mouse globin cDNA to show that these mRNAs are processed to a shorter form after transcription.[15] Why not move beyond globin to study the expression of this much more exciting gene?

On the plan the pair devised together, Lengyel would supply mRNA from an interferon-producing mouse tumor he studied. Weissmann would then make cDNA from this sample and clone it into *E. coli* via pBR322 and G-C tailing. Screening the transformed bacteria would depend on a strategy very like that pursued soon afterward by Pestka: Weissmann's group would immobilize a large pool of different cloned cDNAs on one filter, bathe filters carrying the pooled clones in mRNA from the interferon-producing cell, and inject the retained mRNA into frog eggs for an interferon bioassay. At first a postdoc in Weissman's group, Peter Curtis, tried isolating interferon message from the mRNA sent by Lengyel by separating it in various ways, but by the end of 1977 Weissmann decided this was not working and that they would have to clone cDNA unselectively from mouse tumor mRNA. By forgoing the selective cloning of identified pure genes, this triggered the more stringent biosafety requirements surrounding shotgun cloning at the time (unlike Genentech, with its synthetic insulin, or Ullrich and Seeburg, with their pure cDNA fragments cut from gels at UCSF).[16]

Up to this stage, mouse interferon was the goal and Weissmann had no commercial alliances, but in early 1978 he joined a set of other, mainly European molecular biologists in Biogen, a new company organized by venture capitalists Raymond Schaefer and Dan Adams—the latter from Canadian Nickel (already a major investor in both Cetus and Genentech). Gilbert's agreement to chair Biogen's scientific advisory board convinced Weissmann and the other professors that associating with the company would not cost them too much academic integrity. On joining Biogen, Weissmann immediately was commissioned to clone human interferon. The terms of the early Biogen venture provided that the biologists would conduct company projects in their own labs with company funding for expenses, including postdoc salaries. The professors who were Biogen principals would receive only shares as compensation (which would become liquid assets when the company was launched on the stock market). Biogen would receive exclusive rights to patentable inventions. The professors were free to publish, and the responsibility of asking Biogen to file patents first was left to them.[17]

Clearly these arrangements were designed to help the biologists continue their academic interests and way of life, binding them to the company's prospects only for personal financial reward. They did not, in other words, explicitly constrain research and publication but rather encouraged particular commercially relevant projects. But while Weissmann continued to work on mouse interferon (better in many ways for studying interferon biology), signing with Biogen in March 1978 refocussed his efforts on cloning the human protein. Within days he was on the telephone introducing himself to virologist Kari Cantell in Helsinki, the world's leading supplier of semipure human interferon for both clinical trials and biochemical study—a supply he obtained by processing a substantial percentage of donations to Finland's blood bank. Weissmann persuaded the Finn to collaborate along parallel lines with Lengyel: Cantell would provide the mRNA, which Weissmann would clone, and Cantell would help screen by conducting interferon assays. According to Cantell, Weissmann did not mention his involvement with Biogen then, nor for a long time afterward.[18]

Curtis soon arrived in Helsinki to extract RNA from Cantell's leukocytes as delicately as possible, to take back to Zurich. Because Cantell's normal human white blood cells expressed fewer genes than Lengyel's mouse tumor, they turned out to be a better source for interferon message. Curtis spent the middle months of 1978 again trying to isolate interferon-coding message by extracting bands from gels separating the Helsinki mRNA by size. Given the hazards of RNA handling, it is unsurprising that he failed again. Weissmann again decided on shotgun cloning the cDNA, after only rough sizing with a sucrose gradient. Having laid this groundwork, in the autumn of 1978 Curtis left the lab. Another postdoc, Tada Taniguchi, left shortly afterward, agreeing to leave Weissmann the leukocyte interferon cloning project but claiming fibroblast interferon for himself. Soon Taniguchi succeeded in cloning fibroblast interferon, with funding from Japanese pharmaceutical firms, and in direct competition with Biogen principal Walter Fiers.[19]

At this point Weissmann stepped up the effort by putting three other postdocs on the human interferon project. Weissmann himself came to Helsinki to do the next large-scale RNA extraction with Cantell, around the end of 1978. Postdoc Shige Nagata converted it to cDNA and spliced that cDNA into the *Pst*1 site of pBR322 by G-C tailing, just as Goeddel and Pestka soon would.[20]

Screening and rescreening pools of clones using the elaborate frog egg injection step—and regular air freight to Helsinki—would be arduous and expensive by the standards of 1970s molecular biology, but Biogen was paying the bills. Each screening cycle took a week: 512 different transformed bacterial strains were cultured together, their plasmid DNA extracted and bound to filters, then hybridized with mRNA from Cantell's interferon-producing leukocytes, then washed off, concentrated, and injected into frog eggs. The medium surrounding the frog eggs was then flown to Cantell's lab for interferon assay in cultured human cells. Nagata continued for more than 8 months, testing and retesting 12 pools of 512 clones representing some 6,000 different transformants. Frustratingly, certain pools and subpools gave sometimes positive, sometimes negative results. Nagata believed the positive results, thinking that nothing but interferon sequence could possibly yield a signal, but Weissmann could not believe they had so many genuinely positive interferon transformants, since, like everyone else, he expected interferon to appear in no more than 1 in 1,000 transcripts. They tried some alternatives to the bioassay screening, such as differential hybridization with probe from induced and noninduced leuko-cytes (much like Goeddel and Pestka and the Cetus people). This too yielded inconsistent results.[21]

Meanwhile, pressure was mounting. In early 1979, after less than a year, Bio-gen had burned through its initial investors' money, Adams left, and Gilbert stepped in to act as CEO (as he would for much of the next five years). At mid-year the Germany-based drug company Schering bought a large share of Bio-gen, securing rights to any interferon patents. Pestka, meanwhile, published a piece describing an unprecedented purification of human leukocyte interferon, implying that Roche might be close. In September Taniguchi wrote Weissmann to inform him privately that he had cloned fibroblast interferon. Rumors flew that leukocyte interferon had already been cloned by Peska; by Genentech; by Michel Revel at Pasteur Institute (and affiliated with Cetus); and by Leroy Hood of Caltech, a founder of Amgen in Southern California. Reliable information was impossible to obtain, now that nearly all the established interferon biolo-gists were teamed with commercially backed cloners. Weissmann pushed his postdocs still harder.[22]

Finally one subgroup of 64 transformants registered consistent enough results that it was broken into sub-subgroups of eight clones. The first of these sub-subgroups to give a repeatable positive signal was broken into its eight indi-vidual transformants, and each retested. One positive transformant was soon found, but the cDNA in it turned out to be only about 300 bases long—a third as long as a message for the full interferon protein should be. Just like Pestka and Goeddel, they used that truncated insert as a probe and recovered several other transformants in positive pools with appropriate-sized inserts.

Still these did not consistently test positive in the double bioassay; Weiss-mann needed a clearer way of determining if any of these longer cDNAs con-tained the interferon message. But if the *bacteria themselves* were making active interferon, then their plasmid had to contain the full interferon cDNA, not just a sequence that happened to retain interferon mRNA from leukocytes. First tried as a screening short cut, testing raw bacteria juice for interferon activity was dropped because the *E. coli* extracts proved toxic to the cells used in Cantell's virus protection assay. It was also an unlikely gamble that bacteria could actu-ally produce biologically detectable interferon, even with a perfect cDNA insert, because the main foreign transcript would encode a chimeric protein beginning with half the enzyme conferring ampicillin resistance. For functional interferon protein, the bacteria would presumably have to translate that long mRNA from the middle where the actual interferon coding sequence began, requiring the rare event of bacterial ribosomes binding and commencing protein synthesis at the interferon start codon (or a nearby methionine), without a proper Shine-Dalgarno sequence. And this was not the only worry about whether it would work.[23]

Despite the multiple unlikelihoods, that autumn researchers in Weissman's lab (especially Michel Streuli, a Tufts undergraduate on a study year abroad) developed a simplified cell culture assay for interferon that was less sensitive to bacterial poisoning than Cantell's. With Alan Hall, a new postdoc, the group made a concerted push in December 1979 to test extracts of individual bacterial clones in the positive subgroup with likely sized inserts. On Christmas eve Nagata and Streuli found a clear positive. Weissmann rushed back from an Alpine ski resort, phoned California, and Biogen patent attorney Jim Haley arrived Christmas day. Obviously Biogen felt they had no time to lose. Over the next week, the skeleton lab crew conducted a quick set of other experiments to prove that the source of the antiviral activity in the clone was really interferon. All the while Weissmann and Haley were writing up a patent application covering the cloned interferon-producing DNA sequence, filed in Europe on January 8, 1980—just in time for the Biogen Scientific Board meeting in Martinique on January 12.[24]

At the meeting Gilbert not only approved publication of Weissmann's success without delay but, in order to attract further investment, a fanfare announcement too. To accommodate the convention of first presenting results to scientific peers, the Biogen leadership emulated Genentech's insulin announcement, arranging an informal academic seminar for Weissmann at MIT on January 16, followed immediately by a press conference at a posh Boston hotel. In four days Weissmann managed to write and submit a full journal report, making copies of the draft available for interested journalists and scientists on the day—although this did not prevent scolding for "cloning by press conference." The fanfare sent Schering's stock price up 20% that day (Biogen's was not yet traded).[25]

The many important scientific publications that soon flowed from Weissmann's lab show that despite the hurry, Weissmann achieved a balance between the pursuit of academic biology and the demands of commerce. To mention just some of the noteworthy work done by the group at this time, their March *Nature* report describing their first leukocyte interferon clone as well previously unknown features of the protein was followed by a report in *Science* describing the discovery of a second interferon with a slightly different amino acid sequence and higher activity in human cells, and another *Nature* piece in which the Weissmann and Taniguchi groups collaborated in analysing how closely

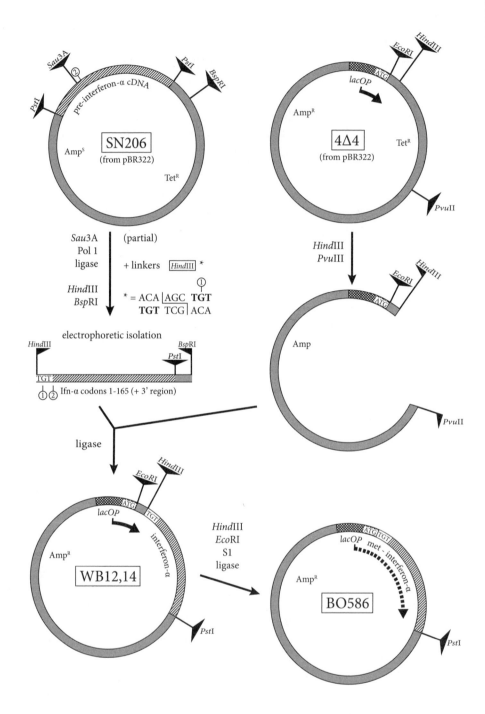

related the DNA sequences of leukocyte and fibroblast interferons were.[26] All of this research cast light on the evolution of eukaryotic genes, a central topic in molecular biology of the day. Even if the same findings about a less glamorous protein might not have gained so many pages in *Science* and *Nature*, this work shows no subordination of scientific priorities.

Meanwhile, back in Weissmann's university lab, Nagata, Hall and the other students labored hard with technician Werner Boll to reclone both the origi- nally isolated leukocyte interferon gene (now called alpha-1) and the second more active one (alpha-2) into newly constructed expression plasmids. This would enable more interferon to be made for academic study of the protein, and also to move toward pilot production and testing of recombinant alpha interferon as a drug. Rather like the Gilbert group's approach to insulin, their main approach was to move the interferon coding parts of the cloned human cDNA, trimmed of their G-C tails and most or all peptide leader by judicious digestion with exo- nucleases and restriction enzymes, to a position where transcription would be driven by a high-expression *lac* promoter. In March 1980 Weissmann's group made a *lac*-alpha-1 construct with a reduced peptide leader that produced a hun- dredfold higher interferon activity in *E. coli* than the original, published clone; by May 1980 they had made another *lac*-alpha-1 construct that produced a thou- sandfold higher activity (despite a small fusion protein stub remaining on the

---

Fig. 4.1. Construction of an expression vector for an alpha interferon 2b cDNA clone produced by Werner Boll and others in Charles Weissmann's group during mid-1980. SN206 was previously obtained (in late 1979) by Shigekazu Nagata from double strand- ed Bowes cell cDNA, with a full-length transcript containing both 5′ and 3′ untranslated portions inserted into the *Pst*I site of pBR322 by G-C tailing. The insert was cleaved by *Sau*3A digestion between the first and second codon of the mature interferon protein, blunt-ended, ligated to a synthetic *Hind*III linker restoring the first codon, and then liberated from SN206 by cleavage by a second restriction enzyme, *Bsp*RI. That cDNA construct was then moved into the *Hind*III site of a *lac* expression vector downstream of an ATG start codon immediately followed by an *Eco*RI site. The sequence intervening between the initiation ATG and first interferon codon was then removed by *Eco*RI and *Hind*III digestion, S1 nuclease treatment, and blunt end ligation. BO586 and derivatives did not produce active interferon for several more months, until most of the coding region was restored from SN206 to repair a secondary mutation. Note that *Bsp*RI and *Pvu*II are blunt-end cutters, and not all *Sau*3A sites are depicted. (Not all enzymes and substrates depicted; see main text for sources and details.) Abbreviations: Amp, ampi- cillin; Ifn, interferon; Pol 1, bacterial DNA polymerase 1; Tet, tetracycline.

hormone). Stymied by a mutation acquired in subcloning, not until October 1980 was the Weissmann group able to make a *lac* construct that expressed high levels of true interferon alpha-2. But eventually they arrived: they were producing a mature human interferon, not a fusion protein, and without relying on Genentech's patent-pending synthetic DNA technique—or not exactly, for they strayed close with a synthetic linker replacing the first codon of mature interferon alpha, and another containing an ATG start codon (fig. 4.1).[27]

Little of this genetic engineering work was ever published, both because its publication prior to patent issue might offer an advantage to Biogen's competitors and also because it was more practical than scientifically innovative. As Hall recalled, "all of the clones I made . . . that wasn't really publishable stuff." Still, the diversion from strictly academic biology led to no great regrets from the academically ambitious Hall and Nagata, and nor in retrospect did their absence from authorship on Weissmann's patent. Both went on to very successful careers in basic research, and felt they had received invaluable training and ample publication credit from their interferon work with Weissmann. Nor is there evidence of conflict from the time; these postdocs experienced less acute tensions than those who left UCSF for Genentech. Perhaps it was as simple as allowing them to claim prominent authorship credit for work like the promptly published discovery, using their cloned cDNAs as probes of genomic DNA, that alpha interferon belonged to family of at least eight different genes. But unfortunately for the scientists of Weissmann's group, they still had to share the limelight: their report of this discovery in an October 1980 issue of *Nature* appeared back-to-back with interferon reports from competitors, who had not been idle. One was Goeddel.[28]

After Goeddel and Pestka were scooped by Weissmann's team in January 1980, the two groups stepped up cooperation and soon retrieved a full-length leukocyte interferon clone of their own. But its sequence differed from the sequence of the Weissmann group's alpha interferon gene, published in June, adding to the mounting evidence that this interferon belonged to a multigene family. In any case Goeddel, Heyneker, and the other Genentech cloners forged ahead with their full-length interferon alpha cDNA to meet the next benchmark in their Roche contract: engineering a construct that could express substantial quantities of the protein.

In building a vector capable of expressing interferon alpha in bacteria, Goeddel and his colleagues essentially followed the same strategy they had just applied to growth hormone. Indeed the expression construct was even tidier and more artful than with growth hormone, and thanks to DNA synthesis more straightforward than the roundabout enzymatic sculpting of cDNA sequence employed by Weissmann's team. Using the same convenient restriction site in the cDNA between the first and second codon of the message for the mature protein (that is, at the second amino acid after the 23-amino acid prehormone signal sequence that would need to be removed anyway) that the Weissmann group would employ, they made and pasted onto the beginning of the interferon sequence a synthetic linker containing an *Eco*RI restriction site, an ATG (methionine) start codon, and a replacement for the first codon of the mature interferon protein. This the Genentech team linked to a high-expression version of the *trp* promoter that they had already created for growth hormone (fig. 4.2). In *E. coli* this construct expressed levels of interferon protein several orders of magnitude higher than the Weissmann group had reported with their fusion protein (and also somewhat higher than Hall and Boll were obtaining at the time in their unpublished work). And while the Weissmann group won most of the scientific credit for finding and describing the human interferon alpha gene family, Goeddel and Genentech filed patents on a number of these variants in June and November 1980. Forced to accept a close second place on both alpha and beta interferon, at least in terms of published results and scientific acclaim, Goeddel reveled in his undisputed victory over competing biologists in mid-1981, by cloning interferon gamma through essentially the same brute force methods based on differential screening (this time preceded by bioassay-guided mRNA enrichment).[29]

Although Cetus and some other early contenders ended up with patents covering some aspects of interferon alpha, Cetus and Amgen both pursuing development of modified alpha products obtained by mixing and matching sequences within the gene family, Biogen and Genentech would emerge with the strongest claims to the natural protein as a first-generation drug. In the US patent office a conflict arose between Biogen's applications on the first interferon alpha genes cloned and expressed in bacteria, licensed to Schering and listing Weissmann as the inventor, and the multiple alphas cloned and claimed by Goeddel in Genentech's applications and licensed to Roche. Schering and Roche quickly made a preemptive agreement to cross-license whatever patents were issued, evidently deciding there was enough money in the drug for the both of them—and too much risk to their own patent positions in litigation. Specific terms of cross-licensing, such as royalty rates, would depend on the results of the Genentech-Biogen patent interference proceeding (a quasi-judicial process conducted by the Patent and Trademark Office). This clash over patent credit between the two biotechs was fiercely contested and lasted a decade, until 1995, confirming the wisdom of the early agreement between the big drug firms.[30]

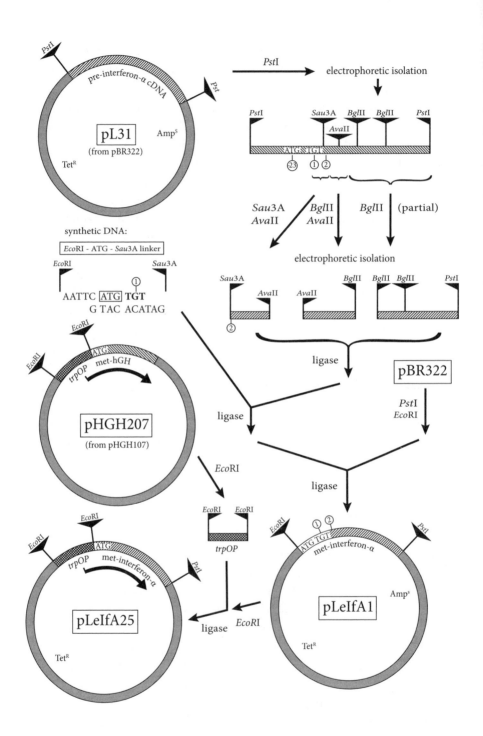

A separate problem arose for Roche and Genentech when the University of California sued them for the unauthorised use of Golde's KG-1 cell line in September 1980. Roche settled with the University in 1982, in part because of bad publicity stirred by a 1981 ABC Television documentary, the *Gene Merchants*. The national program featured Golde, himself yet to be sued by his patient Moore for not asking permission to cultivate and sell his cells, declaring without a touch of irony that "what Hoffmann–La Roche did was not honest. They did not come to the source and ask for the cells but rather went circuitously . . . in secret, and it has a conspiratorial nature to it . . . I felt ripped off." Golde was complaining about a violation of the unwritten code of gift exchange at the time, what Rutter called the "gentlemen's agreement" according to which particular biologists lent each other cells, reagents, and other materials, with the expectation that these would not be passed on to a third party without the source's approval (and inclusion in authorship). Of course Golde was one of the many life scientists whose intimate involvement with commerce was rapidly transforming that genteel code.[31]

The Hopeful Market

Well before interferon was manufactured, tested, and approved for medical sale, the idea of the wonder drug—and all the lesser products expected from the genetic engineering firms—was sold to the public. I refer to the initial public offering (IPO), the sale of equity or shares in a company on an openly traded

---

Fig. 4.2. Expression vector for an alpha interferon cDNA clone produced by David Goeddel and other Genentech researchers in early 1980. Plasmid pL31 was obtained from double stranded Bowes cell cDNA, inserted into the *Pst*I site of pBR322 by G-C tailing. The interferon-coding insert was obtained by *Pst*I cleavage and three contiguous fragments separately isolated by restriction enzyme digestion and electrophoresis: an N-terminal *Sau*3A–*Ava*II segment beginning with the second codon of the mature protein; an *Ava*II–*Bgl*II internal segment; and the *Bgl*II–*Pst*I segment encoding the carboxy terminal of the protein and untranslated 3′ sequences. These three segments were rejoined, and then ligated to a synthetic segment containing an ATG initiation codon and the codon for the first amino acid (TGT) sandwiched between *Eco*RI and *Sau*3A sites, so as to reconstitute the coding sequence preceded by methionine. A high output tryptophan operator-promoter (*trpOP*) terminating in an *Eco*RI site immediately before the *trp* initiation codon, previously constructed to express human growth hormone, was then spliced in to the *Eco*RI site to drive interferon expression. (Not all enzymes and substrates depicted; see US Patent 5,888,808 and main text for further sources and details.) Abbreviations: Amp, ampicillin; Ifn, interferon; Pol 1, bacterial DNA polymerase 1; Tet, tetracycline.

market for the first time. Until the later 1980s, when (other than insulin) their first products began reaching the market, the biotech firms essentially ran on credit alone. Historians of science have observed that the overlapping senses of "credit"—honor for accomplishment, and at the same time belief or trust—have been fundamental to modern science since its seventeenth century beginnings. In the early 1980s, molecular biologists that were able to claim the first two kinds of credit additionally acquired a line of financial credit. The lenders extending it were venture capitalists, and thanks to tax incentives and regulatory changes to encourage private investment during the late Carter and early Reagan years, venture activity in the United States was exploding—from $10 million in 1975 to $4.5 billion in 1983. All this money was looking for somewhere to go, and biotech was a favorite destination.[32]

The IPO is the moment when ownership passes partly from the initial investors to those who buy shares, and the moment the venture capitalist begins taking his profit. Only then does a company come to have anything approaching a concrete value—a price that can be measured against other things openly bought and sold. That value is measured by the dollars that purchasers are willing to pay for shares. If buyers on the stock market pay $10 per share for one million shares, representing half of a company's equity, then the stock market has reckoned the company's overall value at $20 million. The broker who sets the offer price and earns a commission on its sales, the underwriter, has to make an educated guess at what the stock-buying public will pay.

Valuation—equity pricing—prior to an IPO is especially elusive when, as with the biotech firms in the early 1980s, there are essentially no present earnings from which to extrapolate, and no similar public companies with which to compare. For the first biotech companies, estimates of value had to rest mainly on the anticipated future worth of products they would create: mainly medicines. Valuation would have to rest also on the estimated share of a future product's total worth that a given company would be able to claim, through trade secrets and patents (themselves secret until issue, in those days) and know-how: taken together, its scientific assets or simply its "science." And nothing spoke to the value of that proprietary science as much as a company's scientists, especially the eminent ones on the advisory boards. These scientists were the main assets that the public could see; investors from ordinary barbers to Wall Street veterans essentially valued companies according to "who has the best scientists," as the business press noted at the time. And as business historian Robert Teitelman observed, in this haze of massive uncertainty scientists worked in tandem with the stock underwriters—often known as purveyors of market enthusiasm, or "hype"—to support high prices. As one broker put it, speaking of the early biotech offerings, these "stocks are *sold*—they are not *bought*."[33]

At stake in the IPO, of course, was how much wealth would go to early equity holders, including the venture capitalists and many scientists. But for the scientists on the payrolls, and especially for the advisory board members and founders of biotech firms, more was at stake: the IPO measured their value as scientists, because it directly assayed the public's valuation of their science. They all wanted to be able to boast a "multi-million fortune as a clone shark" among their scientific honors, alongside the "piddling $160,000 of the Nobel Prize," as the junior scientists in Gilbert's Harvard department once joked (before Gilbert had either). For them, then, the IPO was what anthropologists have called a "tournament of value," very like the art auctions in which wealthy collectors convert their cash into social status by competitive displays of connoisseurship. Except in this case, the money spent belonged to other people. And the center of attention in this inverse auction was interferon.[34]

The June 1980 Chakrabarty decision to allow patenting of engineered microbes served as a signal for the Patent Office to begin clearing its backlog of biotechnology applications—starting with Cohen and Boyer's patents on basic methods of gene splicing. But well before then, Cetus and Genentech let it be known that they had significant pending claims on interferon and that these were important to company value. Since the patents might not be granted and therefore made public for years, the credibility of their claims on interferon would have to be appraised according to the credibility of its scientists and scientific publications. The way interferon's promise as a treatment for cancer constituted the financial value of genetic engineering was often acknowledged, and perhaps overstated, by biotechnology entrepreneurs at the time. At the Biogen press conference announcing Weissmann's cloning success, a claim to victory in the "interferon derby" (as news coverage put it) underwritten by Gilbert's scientific reputation, Gilbert said that the $2 million being spent on interferon to treat about 100 patients attested the substance's value. Interferon would be "the largest, single most important medical development you and I will ever see," said a Cetus executive in 1980. Whether biotech people actually believed the hype was beside the point; interferon was what gave their companies market value. As Kleid recalled, the interferons were "Genentech's ticket to going public. They were going to be a billion-dollar product, and Genentech had the inside track."[35]

Genentech went to the stock market first that October, a week after the publication of Goeddel's prominent *Nature* piece describing their cloning and expression of alpha interferon. The company could point to revenue-earning arrangements with Lilly, Kabi, and Roche, as well as some prospects for its own products. But in the offering prospectus, interferon occupied more space than growth hormone or insulin or any other prospective products. In the months before the IPO Herbert Boyer, although demure compared to many biologist-entrepreneurs following in his footsteps, played his part with the (other) underwriters and the venture capitalist owners in the obligatory "road show," pitching Genentech stock to banks and other institutional investors with a string of plastic beads to explain gene splicing.[36] The press covered the IPO preparation with unusual interest (so much that regulators delayed the event). Financial reporters observed how remarkable it was for a company with no real earnings to be coming to the stock market in the decade-long atmosphere of financial gloom—especially with one of the largest offerings in history. By way of explanation before and especially after the sensational event, press coverage frequently invoked the new gene technology as a potential "elixir" for America's senescent industrial economy, Boyer as a pioneer in genetic engineering, and of course interferon as its greatest immediate promise.[37]

And a sensation it was. The $35 price set on the million Genentech shares offered—less than 14% of the total equity, thus implying a total company value of $260 million—had at first raised eyebrows, but by October 14, 1980, was thought a little low. As an estimate of market sentiment about this "first chance . . . to bet on the fledgling and esoteric gene-splicing industry," it was far too low: in furious trading the issue shot up into the high $80s on that first day, closing at more than double the offering price. Thus the investing public valued the company at about $530 million, about half of the midsized major drug firm Searle. Wall Street insiders accustomed to valuing stock by "fundamentals" like price/earning ratios began to think that a problematically speculative mood might have settled on the stock market. Again, to the extent that quantitative thinking explained the market's enthusiasm, projection of interferon's use in cancer treatment played a central role. One influential report at the end of 1980 estimated that recombinant DNA products would account for $3 billion in revenue by 1990, half from interferon. Another predicted $3 billion by 1987 from interferon alpha alone. Summing it up, one Wall Street wag observed that interferon had already proven its efficacy as "a substance you rub on stockbrokers."[38]

The lesson of the Genentech spectacle was not lost on Cetus: the public market believed in gene splicing—particularly recombinant interferon—and was ready to bet money on it. Hot on the heels of its then-smaller rival, Cetus announced its own public offering in early December 1980. At $23, the 5.2 million shares sold by the firm—23% of the equity—implied a market value above

$500 million, nearly double what Genentech had slumped to by early 1981. "We wanted the image, I think, to be that the pie is even bigger than it has been represented so far," Cape recalled afterward, as would befit a firm engineering microbes for chemical and energy industries as well as pharmaceuticals. It also befitted a firm with a Scientific Advisory Board boasting such luminaries as Stanley Cohen (Boyer's coinventor on the Stanford gene splicing patents), Nobel-winning biologists Joshua Lederberg, Francis Crick, Andrew Schally, and Hamilton Smith, and also Stanford's Tom Merigan, the famous interferon clinical researcher. These last four had only joined the board in 1980, just in time for the IPO. The prospectus gave top billing to interferon and its Shell-financed codevelopment, among all its medical products.[39]

The huge $120 million Cetus offering of March 1981 did sell, but just barely— the stock closed at its offering price, denying subscribers their customary "pop," or profit from immediate price gain. Although lacklustre by Wall Street standards, the outcome did not prevent Peter Farley, Cetus's physician-president, from gloating that his firm was "the only adequately funded genetic engineering company"—by which he meant valued higher and financed better than Genentech. Cetus's IPO seems to have sent, or at least signalled, a chill wind through the biotechnology share market. One reason offered by analysts was a declining assessment of interferon, although no major new scientific findings could account for this shift. The stock market, in general, had also cooled. By October 1981 many new offerings were being postponed, and Cetus was trading at half its offering price, implying a value at the time about the same as Genentech. The chilly turn was confirmed by the modest September 1982 offering of Genex, the third of the early "Big Four" biotechs to go public (with Biogen, Cetus, and Genentech), whose asking price was scaled back at the eleventh hour to reflect a total company value of only $110 million, despite public claims on alpha interferon and Bristol-Meyers pharmaceuticals as a weighty codevelopment partner.[40]

Unfortunately for Biogen, the tide of money that floated Genentech and Cetus subsided too quickly for their own offering of common shares. Postponing an IPO in 1982, the firm announced in February 1983 both that it would offer 13.5% of the firm for an asking price of $60 million in late March, and that it would commence clinical trials of gamma interferon—to which the firm retained the rights, just like Genentech—in Europe by the year's end. Ultimately the 2.5 million shares were sold at $23, or $80 million in total, reflecting a valuation of the firm at $600 million. Like Cetus's, Biogen's shares barely rose above the offering price on the day. The quiet response of the market was attributed to a more realistic assessment among investors of the timetable for biotechnology revenue, rather than anything negative about Biogen, whose "impressive talent from university laboratories" (comparable to the heavy "scientific talent" behind Genentech's offering) testified to the value of its science.[41] Still, Biogen sensed the market well enough to get their cash infusion, triggering another rush of biotechnology public offerings that spring and summer.

One of these was Applied Molecular Genetics, or Amgen, based in a remote Los Angeles suburb. Established in 1980 by California venture capitalists with a scientific advisory board drawn largely from Caltech and UCLA (and initially UCSF, until William Rutter decided to start his own firm, Chiron, instead), Amgen secured private financing from corporate investors immediately after Genentech's IPO. In the later months of 1981, the firm commenced in-house research on a larger scale than Genentech had at first. The firm's early scientist-managers included Genentech's biology director Nowell Stebbing, previously an established interferon researcher. The main focusses of Amgen's early cloning efforts were a second-generation "consensus interferon" made by mixing sequences from the different interferon alpha genes to optimize its safety and activity profile (capitalizing on Stebbing's expertise), the blood-stimulating hormone erythropoietin (capitalizing on the firm's relationship with the leading American expert on the protein; chap. 5), and combining genes from several microbes to manufacture indigo dye in bacteria. Success was not quick in any project.

Under financial duress, in June 1983 Amgen followed Biogen with an IPO of $40 million in shares accounting for 20% of the firm's equity, modestly implying a value one-third of Biogen's. The prospectus gave top billing to consensus interferon. Amgen's offering fared worse than Biogen's, slumping immediately and signaling to analysts that the briefly renewed spell of market optimism had passed. The prediction was confirmed when Chiron, whose IPO followed Amgen's by less than a month, was only able to find buyers for 1.5 million shares at about two-thirds of their anticipated price, leaving a company valued only at $87 million—and a Rutter very "ungrateful" to the Amgen offering that, he reasoned, had soaked up all remaining demand for biotech investment. The public had implicitly set his firm's value at one-sixth that of Boyer's, one-seventh that of Gilbert's, and still less compared with his Cetus neighbors in Berkeley. (Eight years later Rutter might have had a last laugh when Chiron became the first biotech to turn a real operating profit, thanks to diagnostic products and royalties from the hepatitis B vaccine cloned in his lab and sold by Merck, and subsumed a Cetus ailing from heavy investment in another unsuccessful cancer drug. *Science* magazine's feature on the little-fish-eats-big-fish story credited

Rutter's exceptional scientific leadership for the outcome, making no mention of the earlier unpleasantness around insulin.)[42]

The Market in Hope

The investment market having already assessed the main interferon cloners and taken all bets, recombinant interferon alpha in 1983 entered seriously into the clinical testing that would establish its medical value and enable its sale. All the tests then were small studies. In regulatory-speak, these were Phase I studies on a few healthy subjects to determine the substance's pharmacological profile, and Phase II trials on tens of patients to try out different dosing regimens and to look for effects in patients suffering a particular condition. In Phase II trials, drug effects are assessed by easily measurable biological indicators of "response" linked only indirectly with patient well being, known as surrogate outcome measures. There is usually no comparison group of similar patients receiving a different treatment. Phase III trials, where hundreds of patients with a well defined-condition and health status are randomly assigned either to a new therapy or standard care, and which measure efficacy statistically by a direct measure of clinical benefit like patient survival, are needed to prove whether or not a new drug has value. These would come much later, if at all.

Thus in 1983 the only proven aspect of interferon alpha's clinical value was that, by itself, it did not instantly cure all cancers. Still, news trickling in from Phase II trials had profound impacts on the stock prices of the firms involved, illustrating both the public's continuing expectations of miracle cures and the degree to which the biotech sector was still valued on their basis. For example, Biogen gained 20% in market value when it announced in September 1983 that it would soon start testing its gamma interferon, then lost still more than that ten days later when the *Wall Street Journal* ran a front-page article on the various interferons' current clinical outlook. Far from gloomy, this story mentioned promising, unspecified Phase II results of interferon alpha with Kaposi's sarcoma, that cancerous mark of the new scourge, AIDS, but equally unspecified disappointing results with the common breast and colon cancers (perhaps referring to the British Imperial Cancer Fund's discontinuation of natural interferon trials for these conditions). This story was also apparently responsible for wiping out nearly 25% of Genentech's market value.[43]

The biologists who had done so much, both symbolically and through their concrete actions, to conjure the market forces that would value their science, now surrendered the limelight to the clinical impresarios. These were the influential physician-researchers leading clinical trials, and for interferon the most important were naturally the oncologists—leaders of the specialty charged with what may be medicine's hardest job, caring for people suffering our culture's most terrible disease by instilling hope in all the advanced technology that modern hospitals can offer. The most prominent oncologist with whom Roche worked in early tests of Genentech's alpha interferon was Jordan Gutterman at the M. D. Anderson hospital in Texas. Gutterman had led American Cancer Society and other Phase II trials of natural leukocyte interferon, and was among the first to test recombinant interferon alpha too, as readers of *Time* magazine would have known. In early 1984, Gutterman expressed guarded optimism in an industry periodical about the potential of recombinant interferon for treating several cancers, including kidney, chronic myeloid leukemia, and the rare hairy cell form of leukemia, without reference to any specific evidence. He also opined that the better results he seemed to see in breast cancer with Cantell's natural interferon suggested that cocktails of recombinant interferon alpha subtypes might need to be used for each specific cancer. (His "cocktail" view would have encouraged Amgen, with its "consensus interferon" project.)[44]

Industry-watchers must have read this *Bio/Technology* feature, "Gutterman Talks about the Trials," almost as intelligence leaks. Then, as now, most clinical trials of new drugs were sponsored by their manufacturer and, in the medical literature, reports of industry-sponsored trials are more likely to favor experimental drugs than publicly funded trials. Partly this is because unfavourable reports tend to remain unpublished. That is, by sponsor design, the "good news" in clinical trials gets most of the press. (Although selective publication of sponsored clinical trials has been occurring since the 1930s, the idea that withholding "bad news" could lead to future patients suffering inferior treatments was only beginning to raise serious concern during the 1980s. There now exists a substantial literature quantifying sponsorship-associated bias in the clinical trials literature, in oncology as in other medical fields.) The other leading reason that industry-sponsored randomized trials of new drugs are much more likely to report favorable results than publicly sponsored trials is that drug companies try to sponsor only those tests—especially expensive Phase III trials—that will make their products look good in the first place. They use Phase II exploratory trials to look for health conditions, patient types, and trial procedures that are most likely to show their products effective when scaled up to Phase III.[45]

The years 1984 and 1985 saw a flurry of clinical trials of interferon alpha for hairy cell leukemia, so much so that that it was harder for investigators to recruit subjects than it was for victims of the rare condition—afflicting less than 3,000 Americans out of the 5 million living with cancer diagnoses at the time—to gain access to the eagerly sought drug by volunteering. Why all the industry interest in this tiny market? Gutterman and his Texas colleagues, in a small trial in the early 1980s, had found hairy cell to be a promising therapeutic use for Cantell's natural leukocyte interferon. Identifying a cancer that would show the dramatic level of response necessary to win quick FDA approval without requiring trials involving too many patients, he then focussed on this condition in his testing of the Genentech interferon alpha. The Texas group began a Roche-funded Phase II study with the new drug in January 1984. The results they soon reported were dramatic, with 87% of 30 subjects showing a partial or complete remission based on blood counts and bone marrow pathology. This remarkable figure included some patients who had not yet been treated with the then-standard splenectomy (spleen removal), suggesting that interferon might be suitable as a first-line therapy: the initial treatment for all patients diagnosed with the condition. That idea never entirely caught on, and it soon was realised that many remissions after treatment with interferon were only brief. Still, recurrences of hairy cell leukemia generally could be treated again with interferon.[46]

David Golde, meanwhile, emerged as one of the impresarios leading clinical trials of Biogen's interferon alpha. Following closely in Gutterman's footsteps in early 1984, his UCLA associates formed part of a Schering-funded multicenter Phase II trial of the drug in hairy cell patients who were still sick following the standard splenectomy and chemotherapy (i.e., as a second-line therapy). They were quick to announce preliminary but dramatic response rates in 1985. In 1986 the larger Phase II trial in which the UCLA group participated reported an impressive 89% response rate among 64 subjects.[47]

The FDA approved both Roche's and Schering's recombinant interferon alpha in June 1986 for hairy cell leukemia, its first official use or "indication." Approval for the also-rare Kaposi's sarcoma seen in some people with advanced AIDS was expected soon. Although some in the industry described the wait as long, the FDA had exercised no special regulatory caution. Quite the opposite: the FDA had approved the drug for hairy cell only two years after the relevant trials began, and evidently based only on the type of small trials by Gutterman and Golde described above—that is, without waiting for proof that it improves survival in randomised Phase III studies, as had previously been the rule. Approval came just six months after Roche and Schering submitted applications, and this speed was used by FDA leadership to defend itself from renewed drug industry pressure—now encouraged by the Reagan administration—to be less stringent so as to relieve the "drug lag" purportedly depriving Americans of important new medicines. When it approved interferon alpha for hairy cell, the FDA made explicit what was normally left unsaid and counselled doctors against pre-scribing the new drug for other conditions, noting that it had been tested, and not found effective, against many cancers (never specified, as commercial trial results are held by FDA in confidence).[48] Still, with the drug approved for one indication, no matter how rare, it could be legally sold and used, although not advertised, for any purpose that doctors prescribed—a practice known as "off-label prescribing."

In retrospect, it seems fair to say that recombinant interferon alpha was approved in a hurry, under combined pressure from patients and a general pub-lic wanting to see progress toward a cancer cure, from investors and politicians wanting to help the nascent biotech industry get off the ground, and a Reagan administration and drug industry both pushing to ease regulatory stringency. Interferon alpha can be regarded as the first of a long series of cancer and AIDS drugs introduced to the US market through "accelerated approval." Formalized in 1992, this process, in which approval is granted while still awaiting Phase III proof of a new drug's efficacy, was instituted to help patients access lifesaving new medicines (on the popular but unfounded assumption that new drugs are very likely to be superior to older, proven therapies). But a decade later, more than half of new cancer drugs provisionally approved by the FDA in this way still lacked adequate evidence to confirm clinical efficacy—that is, improvement in survival or quality of life for patients with the cancers they were sold to treat.[49]

Interferon alpha was not much different: in the late 1980s and 1990s, the hope built upon early descriptive trials with recombinant interferon alpha as a treatment for the main cancer killers turned to disappointment. Following up on a promising, promptly published Phase II Roche-funded trial of the Genen-tech product with breast cancer and less promising early studies of their own, Schering funded a Phase II/III comparative trial (carefully controlled but with patient numbers too small to prove anything short of dramatic benefit) in Britain with, and without, the Biogen product after traditional therapy. The results were unfavorable.[50] Gutterman's Texas group also published an exploratory compar-ative study on breast cancer in 1991, conducted some years before, with Roche's Genentech interferon combined with tamoxifen versus tamoxifen alone. Soon after, they also published a 319-patient Phase III study comparing aggressive chemotherapy alone to chemotherapy followed by natural (Cantell) alpha inter-feron. Both showed disappointing results.[51] I have found no published indication that the Texas group conducted a Phase III study to test the efficacy of Genen-tech's interferon alpha in breast cancer, but given Gutterman's candid hints in

1984 that the recombinant product worked less well than the natural product, it presumably would not have reflected well on the drug.

Recombinant alpha interferon was also tested as a lung cancer treatment in the mid-1980s. Initial Phase II studies of the Schering product in Britain and Germany showed evidence of tumor regression in patients treated with recombinant interferon after chemotherapy and radiation therapy, but also revealed serious toxicity ("side effects"). An American group conducted a Phase II/III trial that tested the Roche/Genentech interferon alpha in 132 lung cancer patients responding to initial chemotherapy and radiation therapy and found evidence of benefit—in the minority able to tolerate the very disagreeable high interferon doses for two years. In the early 1990s a European Phase II/III trial compared a total of 219 patients treated with chemotherapy, Cantell's natural interferon alpha or Schering's Biogen interferon alpha, finding no benefit for either interferon.[52] So interferon alpha proved without value in the leading killers among cancers, breast and lung (and colon, where natural interferon's performance was probably too discouraging to attract large trials with the recombinant protein). But this knowledge came years after launch of the products.

Kidney cancer, or renal carcinoma, is another major killer. In 1985, Gutterman's group published results of a Roche-sponsored Phase II study of the Genentech interferon alpha with 56 patients that seemed to confirm responses seen with natural interferon. Since at the time there was no very effective treatment for advanced renal carcinoma, the Texas response rates of 20–30% (measured by pathological and radiological signs of tumor shrinkage, not survival) did look promising. The Texas group followed up with small studies in the mid-1980s comparing different doses of the Roche-Genentech interferon alpha and gamma, with results also said to be encouraging for alpha despite lower response rates and high toxicity.[53] On the basis of such promise Roche sponsored two small trials that, although still mainly exploratory, could have provided indirect evidence that Genentech's interferon alpha did something for renal carcinoma, since they compared the drug to a known, noninterferon treatment. Neither found any major benefit to patients from the interferon beyond established chemotherapy regimens.[54] (Later trials in which the Genentech product was combined with newer chemotherapy regimes showed similar disappointing results and high treatment toxicity.) Still, Roche sponsored a European Phase III trial looking at 178 renal carcinoma patients treated with the Genentech interferon alpha alone or in combination with vinblastine, an older chemotherapy agent (formally, a test not of interferon but of vinblastine plus interferon, since interferon was given to both patient groups on the implicit assumption that it was beneficial). This study found a response rate of 11% with interferon alone, and double that with the combination as measured by the surrogate indicators that defined "response"—but ultimately comparable, low survival rates allowing little room to credit interferon with clinical efficacy.[55] It was with good reason that the FDA required trials proving extension of patients' lives, or improved quality of life, since (as some cancer advocates and AIDS advocates would learn) indices of tumor shrinkage, cell counts, and other surrogate measures are fickle guides to patient benefit.[56]

Lack of FDA approval notwithstanding, and irrespective both of the demonstrated toxicities and the lack of direct evidence that either the Roche or Schering product actually helped patients live longer or better, by end of the 1980s recombinant interferon alpha had become established as a common therapy for renal cell carcinoma. The rationale of the oncologists—challenged as shaky by some among them—was based on averaging response rates in the many small Phase II studies, which at around 15% seemed a little better than older therapies. No (published) Phase III trial to assess recombinant interferon alpha's efficacy formally for renal carcinoma appears to have been conducted until the mid-1990s. Perhaps the drugmakers decided that, given the probably marginal effect of interferon alpha, undesirably large and expensive randomized trials would be needed to demonstrate a statistically significant benefit (if one could be found). Besides, the product was already selling fairly well for such unapproved cancer indications—no doubt aided by oncology's culture of technological optimism, by the plethora of small trials that themselves served as marketing vehicles, as well as by illicit off-label marketing (penalized in the case of Schering's Intron-A). In 1990 global interferon alpha sales reached around $450 million, largely for off-label cancers—certainly far more than could be accounted for by hairy cell leukemia and Kaposi's sarcoma. That made interferon alpha the top selling recombinant drug of the time, ahead of both human insulin ($400m) and growth hormone ($250m). Annual royalties from the roughly $350 million of Roferon and Intron A sales were keeping Genentech and especially Biogen afloat. As one business columnist put it that year, after approval for hairy cell leukemia got the product to clinicians quickly, doctors were "free to experiment with alpha interferon against a variety of other diseases" by simply prescribing it. That the patients ultimately paying for all this experimental prescribing might not have benefited wasn't mentioned. Nor was it acknowledged that qualified investigators with Roche or Schering connections had been "free to experiment" in FDA-registered trials since 1981.[57]

Not that all the hope, hype, and money surrounding interferon alpha failed entirely to benefit victims of common cancers. In 1998 oncologists were seriously entertaining the idea that the high response rate inconsistently seen with the drug in renal carcinoma was both real and due to the placebo effect—the neuroendocrine impact of optimism from treatment with the expected miracle cure. This view was offered by one of the authors of the first published (in 1999) Phase III trial directly assessing the drug's efficacy in renal carcinoma, sponsored by Schering and demonstrating a two-month gain in survival with Intron-A compared with one of the old drugs among a total of 350 patients. Published evidence from comparative trials finally accumulated to a point where, in the early 2000s, advocates of evidence-based medicine could say that interferon alpha does confer a modest survival benefit in renal carcinoma, to be weighed against a significant cost in quality of life.[58] By then a few other cancer uses were approved, but sales of the drug for cancer were overshadowed by its use in hepatitis C, approved by the FDA in 1991. Demand for interferon alpha hepatitis treatment grew into the billions through the mid-2000s, at which point the realization began to spread that, for many patients with mild infections, interferon's toxicity made treatment worse than no treatment. Attention turned to newer antiviral drugs.[59] But by this stage interferon alpha, and the business and medical enthusiasm surrounding it, had played its part in building the enterprise of genetic engineering.

Conclusions

Interferon alpha paid the budding biotech sector's bills throughout the 1980s, first by attracting private investment, then by stimulating IPOs, and in the end through royalties on sales of the drug. This money enabled the bold, blue-sky life science that characterized the work conducted by the early firms. To be sure, there were other factors beside interferon that made the time ripe for biotech ventures, spawning hope, financial and otherwise. To name a few, a general malaise following the defeat in Vietnam compelled Americans to seek novel grounds for optimism, helping bringing Ronald Reagan to power in late 1980 on an upbeat message. There were concerns across the political spectrum about the US loss of postwar preeminence in manufacturing industries, with salvation expected from computing and biotech. And the probusiness federal regulatory regime ushered in by Reagan's election encouraged privatization of public resources including research, and loosened regulation of both pharmaceutical and financial markets. Still, it would not be much of an exaggeration to suggest that without the mutually reinforcing medical and financial enthusiasm around interferon circa 1980, the biotech industry as we know it—a distinct sector based initially in science-driven small firms associated with universities—might never have evolved.

Whether we attribute the intimately linked, inflated values assigned interferon in both financial and biomedical realms as mainly due to overselling or to circumstance, it would be fair to regard the enthusiasms evoked by the protein in the 1980s as a mania. To draw on this wave of unreasonable sentiment in order to fund the science, both in their firms and in their university labs, leading academic biologists accepted a quid pro quo: they helped the financiers drum up investor enthusiasm, and joined with the clinical impresarios stoking medical enthusiasm, in arenas ranging from the IPO prospectus to the pages of national magazines.

Since the race to clone interferon engaged many more biologists, universities, and other institutions with commerce than the earlier insulin affair had done (by 1983 there were at least 26 small biotech firms involved in developing interferon products), it must have had a much broader impact on mores and practices. Due to the inherent demands of commerce, that impact could only have discouraged the old gift economy in materials—a central component of the experimental systems defining many fields of research—within which biomedical scientists like Cantell and Gallo had previously operated, notwithstanding the high ideals about open communication and dedication to basic science prevailing in some leading biotech firms. For as lawsuits like the KG-1 affair quickly showed, it could no longer be normal etiquette to share research materials freely with other researchers, no matter how reputable. Your colleague receiving the materials might be sued, or perhaps even you. Each commercially interested research group had to operate as an island, or better, an intellectual property autarky, reinventing over and again the same wheel that had once been shared by all within a field. The overall cost of this profit-driven duplication of effort has yet to be weighed against the view that, as Boyer asserted in *Time* (next to a picture of a vial of recombinant interferon purportedly worth $250,000): "industry is far more efficient than the university in making use of scientific developments for the public good." But the subjective price in lost cooperation was immediately felt by biologists as a "rupture in the informal traditions of scientific exchange."[60] There was no return to the garden in which the low-hanging fruit had ripened.

CHAPTER FIVE

Epo

The Making of the Biotech Blockbuster

Throughout this book we have explored the active roles university biologists played, during the early days of biotech, in shaping a wider social environment favorable to commercial molecular genetics. In the last chapter we saw how they helped create an overheated investment market for shares in the small biotech firms they founded, and how the hype spilled over into medical overenthusiasm for recombinant interferon use (largely through the marketing clout of the big pharmaceutical partners to biotech firms, to be sure). In this chapter we look at the cloning race that—by capitalizing on a large body of preexisting publicly funded biological knowledge, as usual—brought to market the most profitable, most widely used blockbuster drug ever to come from recombinant DNA biotechnology: erythropoietin, better known as Epo. As we shall see, like interferon alpha, erythropoietin was a valuable contribution to medicine, but was vastly oversold, and this time with catastrophic results. But here we move our focus from how biologists and their corporate allies shaped financial and medical markets to how they shaped the legal reward system for their work.

Specifically we explore the part played by biologists, in their labs and as expert witnesses and as represented by the biotech firms they launched, in the lawsuits that established the patent regime protecting first-generation recombinant drugs. We look at how these lawsuits rebounded on the biologists, through the way the US federal courts defined the boundaries of defensible patents in the Epo lawsuits so as to reward and incentivise certain kinds of commercial molecular biology research and not other kinds, thus setting future research agendas for biotech drug development. We also look at the role of some leading biotech firms, their credibility underwritten by their scientists, founders, and scientific board members, in securing a further government-sponsored monopoly for lucrative recombinant drugs, beyond patents, through the well-meaning Orphan Drug law. Thus in the late 1980s and early 1990s the courts and legislature again became arenas in which biologists strove to impress the public with the value of their work in making drugs through molecular genetics, pleading that their efforts be showered with extraordinary rewards—and in ways that particularly favoured certain factions among them. The ultimate consequences raise questions about whether society might be better served by a different incentive scheme.

Hormone of High Living

As doctors, aviators, and athletes have long appreciated, life at high altitudes increases the body's breathing efficiency. From the end of the nineteenth century, physiologists understood that this adaptation involved an increase in red blood cells and debated whether low oxygen stimulated the production of red blood cells directly or via a hormone. In the 1950s, Yale physiologists were finally able to show that it was a hormone: something in the blood plasma from anemic rabbits made normal rabbits boost their red blood cell counts.[1]

In the Chicago laboratory of medical researcher Leon Jacobsen, a Manhattan Project veteran who studied radiation sickness and recovery with funding from the Atomic Energy Commission and NIH, the circulating hormone was traced to the kidney. One of Jacobsen's students, biochemist Eugene Goldwasser, continued the project of studying this hormone (dubbed erythropoietin, meaning "creator of red blood cells") at Chicago for his entire career. Sourcing large quantities of anemic sheep plasma from Armour Pharmaceuticals, which had been become a major hormone manufacturer in the interwar years by tapping the wastes from the parent company's slaughterhouses, Goldwasser had by the early 1970s become the world's leading Epo expert. Goldwasser determined the molecular weight of sheep erythropoietin and that it was a glycosylated protein—that is, it was decorated with sugar molecules at specific positions along the amino acid chain—but he was not able to purify enough for more detailed chemical or structural analysis. One of Goldwasser's major contributions was to improve assays to measure Epo levels based on the reactions of living animals or cell cultures; these bioassays showed that the protein had effects that differed somewhat depending on the source species and the species in which it was tested, unlike insulin but rather like growth hormone and interferons.[2] Presumably, then, medically useful Epo would have to be derived from humans.

Medical prospects for Epo brightened in mid-1970s with the discovery that the hormone could be extracted from the urine of anemic people. To supply American scientists the NIH sourced substantial quantities of human urinary

Epo from an Argentinian lab, but these quantities were still not enough for Goldwasser to purify sufficient protein for analysis of its amino acid sequence. Then a group of researchers in Japan established a large-scale collection program from anemic hospital patients. One of these researchers, Takaji Miyake of Kumamoto University, brought a large batch of semipurified urine to Chicago for a sabbatical in Goldwasser's lab. In an NIH-funded collaborative project there between 1974 and 1976, he developed a new procedure for purifying the human hormone, which he and Goldwasser published in 1977. After Miyake departed, the same Japanese network became an ongoing supplier for US Epo research. Between 1978 and 1984, under NIH contracts, Goldwasser's lab prepared large quantities of pure Epo from semipurified Japanese urine that, along with lower-purity material from Argentina and elsewhere, was distributed to biomedical investigators deemed deserving by an NIH-appointed committee. Goldwasser himself did not serve on the committee.

Under these arrangements, during the early 1980s all of the high-purity Epo in the United States was prepared by Goldwasser but belonged to NIH. There was no strict accounting of the Epo he produced but informal arrangements evidently allowed Goldwasser to retain some for his own study, since the committee's distribution list did not include him; he also retained some supplies from the batch originally prepared during Miyake's visit. The NIH Epo was used for many purposes beyond the sort of biochemical work Goldwasser conducted, such as experiments on the hormone's biological action in cell culture and animals, the development and clinical use of an antibody-based test sensitive enough to measure Epo blood levels in individual patients, and small-scale clinical trials.[3] Ultimately researchers working on Epo imagined using the hormone clinically to treat severe anemic conditions, the most obvious of which was the chronic anemia that accompanies kidney failure (both because the kidney loses its capacity to produce its own Epo, and because dialysis machines are hard on blood cells). There were about 80,000 patients on dialysis in the US by the mid-1980s, a quarter of whom required such frequent transfusions that they often suffered debilitating iron poisoning, as well as elevated risk from hepatitis (and other blood infections, including AIDS).

The Secret Sequence

At a scientific meeting in June 1980, Goldwasser was approached by UCLA hematologist Winston Salser about collaborating with a new company, Applied Molecular Genetics (soon AMGen). The company was somewhat unusual among start-up firms of the day in that it was headed from the beginning by a drug company executive, Abbott veteran George Rathmann. Salser, as the initial chair of the scientific advisory board, was in the process of recruiting members among California biologists ambitious to emulate the Genentech route to becoming "scientific millionaires." The Amgen scientific board had at first included Bill Rutter, but he soon split from the company to start Chiron. Before Salser himself was quickly replaced, he recruited biochemist Lee Hood at Caltech, who was developing a new kind of amino acid sequencing machine, known as a sequenator, sensitive enough to analyze much smaller protein samples than previously possible.[4]

After some initial discussions with Biogen people (see below), by the third quarter of 1980 Goldwasser was attending meetings of Amgen's scientific advisory board, although he never became a member. Rather, Goldwasser entered into an exclusive consulting arrangement with the new firm, for which he was paid around $10,000 per year—a significant fraction of a professor's salary at the time—plus another $30,000 per year of funding to his lab. To further sweeten the pot, Amgen provided Goldwasser with shares and stock options worth over $390,000 by 1989 (after the company's IPO). The basic plan was that Goldwasser would help Hood in determining the Epo amino acid sequence and otherwise assist the firm in cloning and making the hormone.[5]

Around the beginning of 1981 Hood had obtained some of Goldwasser's high purity Epo protein. Rodney Hewick, a postdoc visiting Caltech from England, ran a sample on the prototype sequenator he had helped Hood design. Hewick had no idea of Epo's perceived commercial potential, but he soon guessed from the excitement generated when he obtained the sequence of the first (N-terminal) 26 amino acids in it. As Hewick recalled, the reaction was "helter skelter; there were people running around and a lot of different people that I hadn't seen before. And, basically, the raw data that I generated was taken away from me." It was given to Amgen scientists, as we shall see. Salser's colleague in hematology at UCLA, David Golde (then collaborating with Genenech), joked to another scientist: "He is the guy that got the amino acid sequence; shall we inject him with something" to make him talk? Cued by such reactions, Hewick wrote the sequence on a piece of paper before surrendering his notebooks and data. Needless to say, the Epo results were not reported in the scientific publication that soon followed, boasting the sequenator's power. Though the 26 amino acids represented only an eighth of the Epo protein's sequence, this unexpectedly available information provided a key to one newly emerging strategy for cloning the gene so that the hormone could be manufactured and sold.[6]

The First Lap

At the time, Epo must have been written on the short lists of many a commercially oriented molecular biologist. Biogen began its Epo cloning effort in 1981, making the protein one of the firm's top priority projects after their interferon triumph. Early in 1980 Goldwasser had been courted by Gilbert to help Biogen, and that June had actually supplied a small amount of his precious, highly purified Epo to Bernard Mach of the University of Geneva, a Biogen principal— but it was not enough to sequence. When Mach asked for more in October 1981 Goldwasser refused (saying that since Biogen had not offered him "any exclusive arrangement," he had made "other arrangements" with Amgen). Thus Biogen's initial cloning strategies did not depend on having Epo protein or sequence. In what can be called the "competition strategy," mRNA was extracted from two sets of the same tissues: one expressing Epo, and one not. Then the mRNA from healthy animals (rats in Geneva; baboons in Cambridge) was used to subtract away the irrelevant messages from the anemic animal mRNA, leaving only the genes expressed specifically in anemia.[7] This could be done several ways. For instance, using a procedure that Weissmann had tried and discarded early in his interferon alpha cloning effort, cDNA could be prepared from normal kidney mRNA, immobilized on a solid surface, and then allowed to mingle with a solution containing anemic kidney mRNA. Messages from the genes expressed in normal kidney tissue would be removed by hybridizing with the immobilized DNA, relatively enriching anemia-specific transcripts in solution. After such a procedure, cDNA clones made from the remaining message would then somehow have to be screened to determine whether they coded for Epo. This would be straightforward enough—provided they had the target gene's sequence. But they did not have it.

Lacking the target gene sequence, Biogen scientists cloned anemia-specific cDNAs and screened them initially by a much more difficult stepwise bioassay-based procedure very like that employed by Nagata in the Weissmann group's successful interferon cloning (chap. 4). Pools of plasmids bearing cloned anemic kidney cDNA could be immobilized on filters and allowed to hybridize with mRNA prepared from anemic kidney. Then after rinsing, the mRNA sticking to the filters was removed and translated into protein in wheat germ extract or a similar cell free system. If a pool of clones included cDNA sequences hybridizing with Epo mRNA, then the cell free system would make Epo protein. To detect Epo protein, product from that translation system would be tested for Epo activity by a biological assay. If any pool of clones tested positive, the pool could be subdivided and reassayed until the one clone specifically retaining Epo mRNA (i.e., an Epo cDNA) was eventually identified. But over about a year spanning 1981 and 1982, Biogen scientists found no Epo clones by this laborious procedure.[8]

Meanwhile, Axel Ullrich at Genentech had also lept into the Epo cloning race, using yet another approach that did not depend on knowing either the gene or the protein sequence: "differential screening." In this strategy cDNA would be made, spliced into plasmids and cloned in bacteria, starting with mRNA extracted from Epo-expressing cells; in Genentech's case, anemic mouse kidneys. To identify clones of the genes expressed only in anemic kidney, as opposed to normal kidneys, Ullrich used separate samples of radioactive mRNA prepared from anemic kidney and normal kidney. These samples were then used as probes on duplicate filter papers bearing spots representing transformed bacteria; clones carrying anemia-specific genes would hybridize only with the anemia RNA probe, but not the normal RNA. This was essentially this same screening procedure used by David Goeddell in his successful cloning of the interferons. (It is unclear how Genentech intended to identify the Epo gene once they had identified the anemia-specific clones; most likely through a bioassay.) Ullrich commenced his work in mid-1981 and continued until the end of 1982, when a false rumor that Amgen had succeeded in cloning Epo convinced Genentech to drop the project. Amgen leadership similarly heard a rumor in 1982 that Biogen had succeeded, but pressed ahead—thanks, evidently, to personal contact between Gilbert and Rathmann.[9]

Biogen ultimately devoted four years and perhaps $6 million in its efforts to clone Epo, engaging about half a dozen PhD scientists and one or two technicians each, at both Geneva and Boston. In 1982 the company switched tactics. Although Biogen scientists still didn't have a DNA sequence specifically from Epo to use as a probe, they had something close: a partial sequence of the protein itself, taken from a slide showing Hewick's results in a late-1981 conference presentation by Goldwasser. (That conference disclosure earned Goldwasser a scolding from Rathmann for his "fundamental mistake in collaboration," but as we will see his candor may accidentally have helped Amgen.) This information leak allowed Biogen to follow the same strategy that Amgen had been using all along: namely, to attempt to reverse engineer the DNA sequence from the protein. This approach was a new technique, and it was difficult. In Biogen's case, it was made more so because the two scientists from the firm in Goldwasser's audience had each hastily scribbled down the string of amino acids on Goldwasser's slide from either end—not quite meeting in the middle.[10]

As noted in chapter 4, there is no one-to-one relationship between the amino acid sequences of proteins and the nucleotide sequences in DNA and RNA that encode them. Only 2 of the 20 amino acids from which proteins are made, methionine and tryptophan, are specified by unique three-nucleotide triplets (AUG and UGG, respectively). Nine are specified by two different triplets each, and the remaining 9 amino acids by yet more triplets. The laws of mathematics dictate, for example, that making a probe to find a gene including the amino acid sequence Histidine-Tyrosine-Lysine—each of which is encoded by two different triplets—would require synthesizing a set of eight ($2 \times 2 \times 2$) different DNA pieces (oligonucleotides or oligos), each 9 bases long, to ensure that one exactly matched. Substitute Leucine, which has six different coding possibilities, for one of these 3 amino acids and the required number rises to a set of 24 different synthetic DNA pieces, each 9 bases long ($2 \times 2 \times 6$). Furthermore, oligonucleotides only 9 bases long are too short to serve as an effective probe; the 1981 paper from Itakura's lab demonstrating this method used 2 separate sets of oligos as probes, each 15 bases long, in cloning a low-abundance cDNA.[11] Thus a researcher hoping to use this strategy would want 2 stretches of protein sequence at least 5 amino acids long, ideally of low enough coding degeneracy to make it possible to synthesize DNA probes for all of the possible sequences.

And longer than 5 might be better, because the same 15-base sequence might occur in a number of places—especially in a very large haystack of sequence that the researchers would be searching for Epo: a genomic DNA "library." Publicly available, the "Maniatis library" created in 1978 by postdocs in Tom Maniatis's Caltech lab (or alternatively the "Lawn library," after the first author of a key publication from Maniatis's group describing it), was a living collection of lambda phage carrying more than a million different 15,000–20,000 base pieces of the human chromosomes. It took about 800,000 phages to achieve 99% certainty of including a given gene. But at least there was that certainty: if one had a complementary probe, one was sure eventually to retrieve a clone of any given gene from a genomic library—provided one was prepared to screen hundreds of thousands of clones, and had a way of distinguishing the true gene from the false positives that coincidentally hybridized with the probe. Thousands or tens of thousands had previously been the rule in screening cDNA clones, but in the case of Epo nobody knew for sure how rare the message was or from what tissue to extract it, making cloning from cDNA very uncertain. The problem was that the longer the probe, the larger the number of different oligos in the probe set tended to be; and the larger the number in the set, the higher the "background" of coincidental hybridization and false positives would be. So to sum up, the best synthetic probe would be fairly long but simple, ideally a unique DNA sequence based on a protein sequence of zero degeneracy, and two separate probes like this were better than one.[12]

For these technical reasons, Hewick's sequence of Epo's first 26 amino acids would have dampened the enthusiasm of any cloner aspiring to use it in 1981, even without Biogen's fuzziness about 3 amino acids in the middle. It contained no fewer than 8 of the dreaded sixfold-degenerate amino acids, and none of the prized nondegenerate tryptophans or methionines. This high redundancy implied a perspiration-intensive slog, fraught with complex synthetic probe mixtures and ambiguous hybridization signals, particularly among the hundreds of thousands of clones in a genomic library. But Biogen forged up this hard path, screening both genomic libraries and cDNA clones with synthetic probe sets— testimony to the keen motivation and high confidence of its scientists, given their known extra burden of uncertainty. Making the task of all its would-be cloners even harder, unknown to anyone the Hewick sequence contained errors at positions 24 and 7. It was a poisoned chalice, and when published in 1983, it lured still others to its hazards.[13]

Biogen scientists were not undaunted. Hoping for more and better protein sequence for purposes of probe design, Biogen's Bernard Mach tried to sidestep Goldwasser by applying for pure Epo from the NIH committee that distributed it for research. In February 1982 the NIH committee declined Mach's request because, they said in confidential communication between themselves, "Gene [Goldwasser] would not have appreciated it," since he had previously turned down collaborating with Biogen. If the decisions of this NIH committee can be taken as representative, the American life science community must have adopted the attitude that the academic scientists who had devoted the most effort to particular research projects with public funding deserved first crack at commercializing them. Certainly Biogen had no further luck obtaining more of Goldwasser's high-purity Epo that, as the firm's Boston lab director Richard Flavell aptly put it, "was like liquid gold . . . treated by the NIH like the Crown jewels."[14]

Amgen, meanwhile, had been working on Epo since January 1981, even before establishing its own labs at the end of the year. Starting that month protein chemist Carlton Paul, Amgen's first scientific employee, tried to confirm Hewick's sequence and obtain more from Goldwasser's high-purity Epo protein, using Hood's sequenator and leased space at Caltech. In August 1981, molecular geneticist Fu-Kuen Lin became approximately the firm's seventh scientist, having left a junior faculty position in South Carolina where he had just cloned duck globin from a genomic library using mRNA as a probe. Lin, joined by a technician in December, took on cloning Epo as his main project, and Amgen's Scientific Advisory Board expected success by mid-1982. Lin explored several avenues to cloning Epo, such as synthesizing pieces of Epo protein from the known sequence in order to produce strong antibodies (which might be used either to test for Epo production from clones, or to clone the gene from cDNA by capturing mRNA attached to ribosomes in the process of producing Epo).[15] But his main cloning effort involved using synthetic probes derived from protein sequence to screen the Maniatis genomic library.

Lin's first such probe set in October 1981 was a single long "20-mer," that is a string of 20 nucleic acids, spanning amino acids 20–26 at the inside end of the Hewick sequence. This first batch, prepared for him by Colorado professor Marvin Caruthers (Goeddel's old advisor, now an Amgen principal), contained 48 different 20-mers, representing only a quarter of the "fully degenerate" set of 192 needed to fully cover all possible combinations for the target sequence. He soon ordered more probes, but these were not fully degenerate probe sets either, with one exception: a fully degenerate 32-member set of 14-mers from the 5 amino acids in positions 18–22 (table 5.1). This probe set, obtained in January 1982, must have discouraged him with its very high background noise. Fighting the background problem, for the rest of the year Lin tinkered with hybridization conditions while relying on "guessmer" probes, small subsets of the full set needed to cover all possible nucleotide sequences near this same low-redundancy stretch of amino acids 20–26. No combination of two probe sets yielded a clone that hybridized with both, and could be confirmed as encoding known Epo protein sequence. Given the Hewick sequence's error at position 24, his continuing failure comes as no surprise. By the end of 1982 Amgen CEO George Rathmann and research director Daniel Vapnek rated the Epo project as a "disappointment" and even "dead"—unfruitful compared with projects started much later. But Lin was granted another chance. In early 1983 he began looking into new avenues to clone Epo and, in a strategy rare for molecular biologists, reflectively testing his screening method to explain its failure in a model system (a gene he had already cloned).[16] Meanwhile Amgen's executives and scientific managers tried to wring more pure Epo from Goldwasser, to obtain new protein to sequence for probe design.

By now Amgen had yet another formidable competitor in its quest to clone Epo. This one emerged in the second half of 1979 from Harvard University's investment managers, around the same time that the four venture capitalists that started Amgen began laying their own plans. Harvard's President Derek Bok, displeased by the way that Gilbert pursued outside business arrangements

TABLE 5.1

Characteristics of probes sets used for screening by FK Lin at Amgen and Edward Fritsch at Genetics Institute, by date ordered (and used soon afterward for screening, except when otherwise noted). In both Lin's and Fritsch's case, the second group of screening efforts reflects the new availability of Epo protein supplies sufficient to design new probes internal to the amino-terminal "Hewick sequence." AA position refers to the amino acid sequence used to design nucleotide probe sets, as counted from the amino terminal of the mature protein.

Scientist	Date Ordered	Epo AA positions	Size	Oligomers/ full set	Fully degenerate?
Lin	Sept-81	20–26	20mer	48/192	no
Lin	Jan-82	18–22	14mer	32/32	yes
Lin	Feb-82	20–26	20mer	48/192	no
Lin	Feb-82	20–26	20mer	16/192	no
Lin	Mar-82	20–26	20mer	8/192	no
Lin	Jul-82	20–26	20mer	1/192	no
Lin	Jul-82	19–25	20mer	4/256	no
Fritsch	Sept-82	18–24	19mer	256/256	yes
Fritsch	Sept-82	22–26	14mer	48/48	yes
Fritsch	Sept-82	18–22	14mer	32/32	yes
*Lin**	*Apr-83*	*62–67*	*17mer*	*64/64*	*yes*
Lin	Sept-83	86–91	17mer	128/128	yes
Lin	Sept-83	46–52	20mer	128/128	yes
Lin	Sept-83	18–23	17mer	128/128	yes
Fritsch	Apr-84	144–150	14mer	96/96	yes
Fritsch	Apr-84	46–52	17mer	32/32	yes
Fritsch	Apr-84	46–52	17mer	128/128	yes

* Lin's April 1983 probe set was never used for screening due to low specificity

while still a professor using university facilities, wanted to make Harvard a major investor in its own biotech venture. By early 1980 Bok and the investment managers had convinced molecular biologist Mark Ptashne to organize the scientific program. The company was to be called the "Genetics Institute" (GI) to communicate the academic-style corporate culture then thought to attract the best young molecular biologists—and to avoid anything with "tech" in it, so as not to raise alarms among Cambridge's biotech-leery populace. But the Harvard Arts and Sciences professoriat reacted badly when Bok informed them of these developments in October 1980. Some academics were fearful that basic research would be diverted and faculty hired based on their profitability to the university. More entrepreneurial faculty had their own objections. Gilbert said he would "resent" having to pit his "own company" against his employer's, and other faculty already involved with industry also feared that Harvard might favor the scientists working for the university's own firm. Bok dropped the plan after a prominent debate that autumn of 1980, when these incompatible notions of "conflicts of interest" converged against it. With a new conflict-of-interest policy put in place in 1981, Gilbert resigned his Harvard post in 1982 to serve as Biogen's CEO for the next three years.[17]

By this time the Genetics Institute project had taken on a life of its own. Ptashne and some of the other people already recruited to the project decided to continue without Harvard, finding venture capitalists to substitute for the university's stake. Ptashne's first scientific board recruit and company cofounder was Tom Maniatis, formerly a postdoc in his lab, who returned to the Harvard faculty from Caltech in 1981. One of the first scientists GI hired, in April 1981, was Rodney Hewick—the man with the Epo sequence, as Maniatis had to know from Caltech colleagues and from Golde, now also part of the GI circle. Soon after Hewick became the firm's lead protein chemist, he told the GI people everything he knew about the 26 amino acid sequence he had copied down while at Caltech, including "what was solid and what wasn't" (i.e., where amino acid identification was uncertain). And with Hewick Ed Fritsch, onetime Caltech postdoc and coinventor with Maniatis of the human genomic library, had by the end of 1981 agreed to join GI. Thus the firm could claim two of the three scientists who literally wrote the book on cloning: *Molecular Cloning: A Laboratory Manual*, by Maniatis, Fritsch, and Joseph Sambrook, the big blue spiral-bound "cloning bible" from the Cold Spring Harbor summer cloning course that Maniatis taught in 1980 and 1981, and that from 1982 graced the shelves of virtually every molecular biology lab in the world. So at the outset, GI might have been a bettor's favorite in the race to clone Epo.[18]

Fritsch officially started at GI in April 1982 but had other projects besides Epo. Also, he may have forseen problems with the Hewick sequence's paucity of low-degeneracy targets, because in 1982 GI approached Miyake—Goldwasser's original source—for a quantity of pure Epo in order to determine more protein sequences for probe construction. But the price was too high (Fritsch recalls more than half a million dollars, perhaps 20% of GI's available funds), so they used what sequence information they had. In September 1982 Fritsch ordered three fully redundant synthetic DNA probe sets from the Hewick sequence, all from the segment between amino acids 18 and 26 (GI was spared Biogen's problems with the error at position 7, because Hewick doubted his identification of amino acids there and at position 3). In October Fritsch screened his genomic phage library with two of them, becoming the first among the competitors to use two fully redundant probe sets, a matter significant in later court action. These were a set of 256 19-mers, from amino acid positions 18–24, and a set of 48 14-mers, from positions 22–26 (table 5.1). He obtained a few clones that hybridized with both probe sets, as expected for the true Epo gene. However, none turned out to be the Epo gene—unsurprising given the error at position 24. He had no better luck in January 1983 with a third fully redundant probe set, even though it came from error-free amino acids 18–22, because his tests to identify the positive clones by a second hybridization involved one of the other two erroneous probes. Thus by early 1983 GI was as stumped as the Biogen and Amgen teams using almost the same approach. Fritsch turned much of his attention to other projects for the rest of 1983.[19]

Back to the Source

At the end of 1982, with Amgen's effort, like everyone else's, mired in the hazards of the Hewick sequence, product development chief Nowell Stebbing (recruited from Genentech) decided that what the firm needed was yet more Epo. The actual quantities of pure Epo to be supplied by Goldwasser were never specified in his agreement with Amgen. Goldwasser had not been completely unforthcoming. Goldwasser's 1982 shipments to the lab had included a little more pure Epo in April, and in June a set of 15 protein fragments made with the standard protein-cleaving enzyme trypsin, separated and purified in his Chicago lab. Neither yielded protein sequence useful for designing new probes. So in December Stebbing sent wrote Goldwasser a stern letter reminding the Chicago professor that Amgen expected that the $30,000 annually to his lab "would be directed toward a focussed effort to provide reagents which are critical to our program," especially "purified Epo."[20]

Around the end of February 1983 Goldwasser obliged by sending another batch of Epo tryptic fragments. These were sequenced by Por Lai, Paul's replacement as Amgen's protein chemist. Evidently wary of repeating 1982's fruitless cloning efforts, Lin tested some new probes based on sequence from these fragments by checking if they would hybridize preferentially with any band on a gel separating anemic monkey kidney mRNA by size (a "Northern blot"). When the probes failed this Northern test, Lin decided not to use them for screening. Amgen seemed back at the drawing board.[21]

Amgen's CEO, Rathmann, later recalled that the company's "ultimate frustration" was that quantities of high-purity Epo protein supplied by Goldwasser were "just plain not enough." As Rathmann understood the situation, Goldwasser had "very good reason to hold back" because his academic career depended on his own supply of high-purity Epo. In July or August 1983 Rathmann sent a delegation, consisting of the company's research director Daniel Vapnek and biologist Joan Egrie, to Goldwasser's lab in Chicago. On arrival they quickly established by "perusing his place" that Goldwasser in fact "had a lot of the material, and he was storing it very fastidiously." The next day Vapnek and Egrie "loosened him up;" Goldwasser, evidently convinced that Amgen really needed the protein and would not waste it, soon sent a fresh batch of high-purity Epo fragmented by trypsin. By September 2, 1983, within a few days of its arrival, Lai had sequenced the fresh Epo fragments, allowing Lin to request fresh synthetic probe based on the new sequence information, a fully degenerate set covering amino acids 46–52. Two weeks later Lin requested another set from amino acids 86–91, again based on Lai's new sequence information. This probe set picked out a band from a gel blot made with anemic monkey kidney mRNA under conditions where it did not hybridize to a nonanemic blot, passing the Northern test. By the second week of October Lin had screened his genomic library and recovered 24 clones that hybridized strongly with both probe sets, and by the end of the month chose four of these for DNA sequencing. In early November, DNA sequencing confirmed that two of the genomic clones agreed with known Epo protein sequences. On December 13, 1983, Amgen filed a patent on DNA encoding Epo in Lin's name, and soon after issued a vaguely worded announcement that it had successfully cloned the gene.[22]

Thus, a mere two months after Goldwasser's more generous protein shipment, Amgen had human genomic Epo clones, and also a cDNA Epo clone from anemic monkeys. In barely three months they had their cloning confirmed and their patent filed. It was Goldwasser's protein shipment that revived the "dead" project, not any major change in cloning procedure at Amgen. All told, Amgen received 20,000 picomoles of high-purity Epo directly from Goldwasser during the relevant two-and-a-half years between 1981 and 1983, a period during which the NIH committee was distributing 100,000 picomoles per year.[23] In other words Amgen was consuming 5–10% of the US supply of purified Epo, and relatively cheaply too—testimony to the magnitude of Amgen's dependence on Goldwasser, and by extension publicly funded life science, for that firm's cloning success over rivals.

Reflecting the corporate culture Rathmann had brought with him from Abbott, where he had acquired the attitude that the firm benefited if scientists were free to follow their curiosity, but could only be harmed by free publication, the Amgen team did not promptly claim scientific credit with a high profile journal article. Rather, the first quarter of 1984 saw Amgen product development teams moving to "get a first generation Epo product into clinical trials as soon as possible." Unable to obtain a human cDNA clone quickly enough, the firm simultaneously pursued four alternative pathways to produce the hor-

mone: semisynthetic vectors designed for expressing Epo in both yeast and in
E. coli; expression of monkey Epo in yeast using the monkey cDNA; and expression of the full-length human genomic Epo in cultured mammalian cells (in
particular, Chinese Hamster Ovary [CHO] cells). This last pathway offered the
advantage that the product would carry sugar residues at particular points along
the protein chain similar to the natural human hormone, because animal cells
had the capacity to add them, but it depended on the cultured hamster cells correctly making and splicing transcripts from the full-length human gene. The
other pathways were backup plans. In the event, Amgen scientists were lucky
and the hamster cells read the human genomic DNA correctly. By the fourth
quarter of 1984 the firm was moving toward pilot production with a CHO cell
line engineered with a vector not unlike that constructed but not used by Berg
in 1972, called pDVSL: it contained pieces of pBR322 enabling its survival in bacteria, pieces of the monkey virus SV40 genome for survival in animal cells, an
SV40 promoter to drive expression from the human Epo gene, and in addition
pieces of mouse DNA carrying the dydrofolate reductase gene (DHFR), which
detoxifies a chemotherapy agent (fig. 5.1). When transformed with the plasmid
and grown with that agent, surviving hamster cells stably integrated the foreign
DNA in their chromosomes, and some produced Epo protein in quantity.[24]

The scientists at Biogen, too, had determined that their Epo cloning effort
depended on access to more of the protein. They attempted to acquire it from
well-known academic researchers in the field, but only received samples too
small to sequence. For some time in 1982 into 1983, the Biogen team in Geneva
collected thousands of liters of urine from hospitalised anemia patients, trying to purify and sequence the protein themselves. This effort went badly awry:
the Biogen scientists did purify an anemia-specific protein, obtained an amino
acid sequence and designed synthetic probes from it, and then succeeded in
obtaining clones from their anemic rat kidney cDNA libraries that matched the
probes. But the protein was not erythropoietin (as Geneva lab chief Julian Davies
quipped in court, "we got some nice genes" anyway). At this point, understandably, they switched their emphasis to screening genomic libraries. In late 1983
Biogen purchased a large quantity of partly purified urinary Epo at great expense
from Green Cross, one of the largest producers of blood products in Japan. Protein chemists at Biogen's Boston labs were able to sequence the first section of
the unfragmented hormone, but they came up with the same sequence that
Hewick had. In early 1984 they received another batch of Epo from Green Cross
that finally yielded some internal sequences useful for probes. But these probes
also failed to retrieve genomic clones. Biogen scientists finally succeeded in

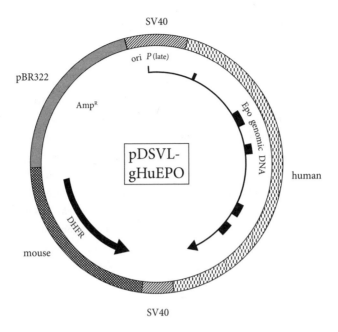

Fig. 5.1. The vector constructed by Amgen around the end of 1983 to transform Chinese hamster ovary cells so that they would stably integrate and express erythropoietin (Epo) from the human genomic clone obtained by FK Lin through the SV40 late promoter. (See US patent 4,703,008 and main text for further sources and details.) Abbreviations: Amp, ampicillin resistance; DHFR, dihydrofolate reductase; ori, replication origin.

cloning the Epo gene in 1985, first a baboon cDNA and then, using that clone as probe, the human sequence from as genomic library, but it was too late and the company dropped the project.[25]

Back at Genetics Institute, Fritsch and collaborators had applied the dual probe set strategy successfully to cloning the gene for clotting Factor VIII, a blood protein deficient in many cases of hemophilia. GI filed a patent application on Factor VIII in October 1983 and was ultimately granted broad claims covering the use of fully degenerate probe sets to clone from a genomic library—essentially the same technique that Lin, whose patent application came in December 1983, had used to clone Epo. Near the end of 1983 Fritsch returned his full attention to cloning Epo. The Factor VIII success suggested that the problem had lain not in the method but in the Hewick sequence on which probe design had been based. In exchange for Japanese rights, the firm contracted with the Japanese pharmaceutical firm Chugai to help bankroll the Epo effort (while Amgen similarly obtained backing from Chugai's competitor Kirin). In January 1984 GI reopened negotiations with Miyake, now working at Wright State University in Ohio, for high-purity Epo protein in order to obtain internal sequence for more probes. This time they agreed on a price GI could afford.[26]

However, news of Amgen's success posed a problem for GI: their new Japanese partner wanted to back out. GI's arguments to Chugai are worth reviewing. In short, they pointed to the vagueness of their rival's announcement (without a full scientific publication) as a reason to keep working, since it was uncertain what, exactly, Amgen had cloned and what patents it had applied for—let alone which claims would actually be granted. Furthermore, even if Amgen had a real Epo clone in hand, with GI's expertise in expressing cloned genes in mammal cells and protein purification, they might still beat Amgen to market with a drug. Chugai found these arguments convincing, even though the possibility of future patent litigation was obvious. The company must have considered that the benefits to be weighed against the risks were large—a "mountain of treasure" from the blood-boosting hormone, as Chugai executives described it a year later.[27] Clearly, by this stage of the biotechnology revolution, scientists understood as well as lawyers that patent priority did not guarantee marketplace exclusivity; rather, as the old saying goes, patents only provide the keys to the courthouse.

The GI group plunged into cloning the hormone with great urgency in early 1984. In April Hewick received his first shipment of high purity Epo from Miayke. Though there would be three more batches by November, the sequence Hewick got from this first batch sufficed for the cloners. Synthetic probe sets were made from the protein sequences of positions 46–52 and 144–150, the former a tryptic fragment also used by Lin. By the end of May Fritsch's team had retrieved human genomic clones, and, after confirming that the cloned DNA matched the Epo protein sequence, in July found cDNA clones in a library that Fritsch had made from human fetal liver. By August GI confirmed that they had several clones containing human Epo sequence, and in two or three months moved a human cDNA into CHO cells. Three months later near the end of 1984 GI filed a US patent; a month later the firm also had lodged patents in Europe as well as Japan, both jurisdictions where patents are preempted by publication. Fritsch then sent a report of the success to *Nature* that was published in February 1985, securing much of the scientific credit (as Lin's work remained unpublished until later that year). For most of 1985 GI worked on improving their Epo cDNA-expressing CHO cells, made with a construct rather like Amgen's but with refinements such as deliberately introduced enhancers, translation stimulators, and terminators, all recently discovered gene expression elements unique

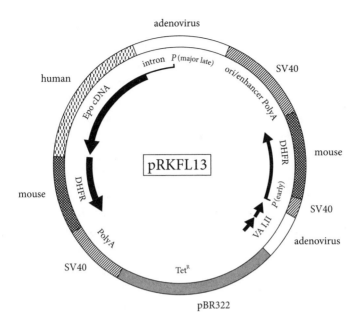

Fig. 5.2. The vector constructed by Genetics Institute in 1984 to transform Chinese hamster ovary cells so that they would stably integrate and express erythropoietin (Epo) from the human cDNA clone obtained by Edward Fritsch and colleagues, through the Adenovirus major late promoter. (See European Patent 411,678 B2 and main text for further sources and details.) DHFR, dihydrofolate reductase; ori, replication origin; P, promoter; Tet, tetracycline resistance; VA, virus-associated transcripts (translation enhancers).

to eukaryotes (fig. 5.2). Soon important studies on the regulatory regions of the gene in humans were conducted with the commercial genomic clones, GI giving theirs to several academic groups and Amgen's in Goldwasser's lab.[28]

Meanwhile, Hewick had made a surprising discovery about Miyake's "pure" Epo protein—which was prepared from anemic urine in basically the same way Miyake had developed with Goldwasser in 1976–1977. Even though it looked like one band on Miyake's and Goldwasser's protein gels, when Hewick processed Miyake's first large Epo batch of April 1984 by the relatively new procedure of reverse phase high pressure liquid chromatography (RP-HPLC, employing a hydrophobic stationary phase rather than the traditional hydrophobic solvent), he found that the preparation actually was half made up of other proteins. Working with the ongoing supply from Miyake through the rest of 1984, Hewick was able to improve and scale up purification procedures in GI's state-of-the-art protein lab, and was able to obtain a more homogeneous protein preparation with double the specific Epo activity (that is, biological activity per gram) achieved by Miyake and Goldwasser. GI applied for a US patent on Hewick's improved purification procedure—and its product, pure "homogeneous" Epo—in January 1985. Though partially revealed in the 1985 *Nature* article, no publication on the procedure was submitted to a journal until late June 1987 when the patent was issued. To competitors, the patent was a nasty surprise.[29]

Policy by Other Means

If war is politics by other means, to paraphrase Clauswitz, then the legal war over the Epo "mountain of treasure" became science policy by other means. Conflict erupted on the very day in late October 1987 that the US Patent Office granted Lin's Amgen patent on the Epo gene. Prepared for battle, Amgen immediately filed infringement suits against Genetics Institute and Chugai. Two days later GI filed a counter-suit against Amgen for infringing Hewick's patent on purified Epo protein, which had been granted in June. Named also in Amgen's suit was Integrated Genetics of Massachusetts, which had used GI's published sequence to clone their own human cDNA in 1985 and then developed its own expression system for production from cultured cells. There were other competitors too, such as biologist Jerry Powell of the University of Washington, connected with Seattle startup Zymogenetics, who distinguished himself among all cloners by successfully retrieving human Epo from a genomic library with synthetic probes based only on the Hewick sequence. But for Amgen, just as for Genentech in its almost simultaneous struggle to lay claim to tissue plasminogen activator or tPA (chap. 6), GI was the competition that mattered—because of its credibility and competence, and because its product was so quickly approaching market.[30] If Amgen would not share the market, GI had to be beaten, or at least delayed.

By this time Genetics Institute had clinical trials of its Epo underway in Japan, through Chugai, and in Europe through its partner there, Boehringer Mannheim. Amgen had completed a small (Phase II) clinical trial in the US using Epo to treat anemia in kidney dialysis patients, which made headlines for improving quality of life in mid-1987 (although in fact the studies only proved that Epo increased blood count). Having two years earlier sold US rights to market erythropoietin for other anemias to drug giant Johnson & Johnson, in a problematic contract that would lead to sales competition and expensive litigation between the erstwhile partners, Amgen was now pushing to receive FDA approval to sell the drug for kidney failure patients in the US. This market was then estimated to be worth $150 million annually, or for the estimated 20,000 dialysis patients receiving frequent transfusions, a very high $7,500 per patient. Amgen had a 300-patient comparative Phase III US trial well underway to prove efficacy—that is, to establish whether Epo could reduce transfusions for dialysis patients without too many countervailing side effects. Other competitors also had commenced clinical testing, including Integrated Genetics and also a firm called Elanex, with a first generation Epo drug based on Powell's clones.[31]

Amgen and GI, however, held by far the strongest intellectual property claims. At stake in their patent infringement litigation was not just which company would survive and which investors would profit, but ultimately the manner in which patent law would motivate investment and shape biotechnology drug development efforts in the United States. There were many areas of legal uncertainty, helping to explain the investment in research and development of Epo by so many rivals (over $16 million by GI alone). Nobody yet had tested how the courts would support patent claims to entities and processes developed through the application of molecular biology. And magnifying the unpredictability, an entirely new American court was established in 1983 to hear patent litigation appeals, the Court of Appeals for the Federal Circuit (CAFC).[32]

The dividing line between invention and scientific discovery, that is between patentable contributions to manufacture and unpatentable products and laws of nature, was unclear when it came to naturally occurring DNA sequences and proteins. However there was good reason in the 1980s to suppose that the courts in the US would be generous to cloners on the issue of subject matter. In a body of case law largely built up around the pharmaceutical industry's commercialization of hormones in the early twentieth century, naturally occurring substances *as isolated and purified* could be patented if they were useful in ways that their natural source was not. Nevertheless, nobody knew for sure what scope would be afforded patents on cloned natural genes. A common expectation that the scope would be broad enough at least to provide a monopoly on first-generation, unmodified protein drugs had just been destabilized by the British High Court's 1987 ruling to the contrary in the important Wellcome-Genentech litigation over tPA (chap. 6). And even before this, firms that openly cloned human Epo based on GI's published sequence and then developed their own expression systems, like Integrated Genetics, were betting on a narrower scope.[33]

Beyond the "product of nature" issue, a patentable invention must be "new," "useful," and "nonobvious." The question of what US courts would accept as a "nonobvious" invention was particularly hard to predict in the mid-1980s, given that most of the techniques in use in biotechnology represented incremental refinements in the shared methods of molecular genetics or protein biochemistry. The British High Court had rejected Genenetch's broad tPA claims on just these grounds of methodological obviousness (chap. 6). It was fair to say that neither Hewick's protein patent nor Lin's synthetic probe method disclosed any earth-shakingly new techniques by the standards of scientific journals (the most prominent of which, like *Science* and *Nature,* devoted far more attention to the natural characteristics of Epo and Epo genes than the methods used to isolate them). As noted, Itakura and colleagues at City of Hope had invented the basic cloning strategy. And neither leading stock analysts nor Amgen's corporate partner in commercializing Epo, Johnson & Johnson, regarded Amgen's patent claims on the naturally occurring (genomic) Epo sequence as clearly superior to GI's claims on the equally natural pure protein. Rather, the financial and legal experts considered it advisable for both firms to negotiate cross-licensing, allowing each to commercialize the drug and to exclude rivals like Integrated Genetics without running the risk that their own patent (or both) would be thrown out by the courts—just as Roche and Schering had done with alpha interferon (chap. 4). GI was amenable to a pact, but Amgen refused.[34] The Federal courts would have to decide how high to set the innovation bar in molecular biology required to win the lucrative monopoly that a patent granted. Amgen was effectively betting that the courts would handsomely reward the first company to reach the market with a new biotech drug, even if based on marginally innovative ways of isolating natural entities—and that if it came to a tie, that DNA would trump protein.

The legal battles were lengthy, expensive, and unusual for the broad set of arenas they occupied. One was the International Trade Commission, where in January 1988 Amgen filed suit to prevent Chugai and GI from importing Epo (for experimental work) made with GI's allegedly infringing cDNA. Over the next two years, the ITC rejected Amgen's complaints several times on the basis that the Lin patent covered only Epo DNA and not any process for manufacturing Epo protein—an unsurprising ruling, since the US patent office had forced Amgen to strip away such claims because they described insufficiently innovative processes. (The issue was crucial in the ITC because that case centered on the claim that GI violated tariff laws specifically forbidding import of foreign goods made by *processes* infringing US patent-protected manufacturing processes.) In January 1989, on the heels of an important ITC decision to this same effect, the Massachusetts Federal district court made a preliminary ruling that Amgen infringed the GI Hewick patent, but did not throw out Amgen's Lin patent either (as GI had moved). To expedite a full decision on the validity and scope of both GI's and Amgen's patents covering Epo, both firms agreed to a trial before a magistrate. During the trial, in June 1989, the FDA gave Amgen approval to market its Epo product Epogen for anemia in kidney failure.[35]

In December 1989 the Boston magistrate Patti Saris decided on the many questions at issue. That the court found the decision "close" is no surprise, given both the closeness of the cloning race—in which Amgen first isolated the genomic sequence, but GI the human cDNA—and the distinction of the scientific principals and expert witnesses called by both sides, including Hood and Maniatis, as well as Ullrich, Flavell, and Davies. On the product of nature issue, Saris took the standard line that Lin's patent applied to the isolated Epo gene, not the gene as it exists in the body. On the issue of scope, Saris upheld most of the claims of Lin's patent, such that the natural genomic sequence also covered the human cDNA sequence, but excluded claims to every possible (unspecified) DNA sequence that encoded any protein with erythrocyte-stimulating activity. Another thorny issue was date of conception. On this, Saris ruled against GI, denying that Fritsch having first demonstrably planned to use dual fully degenerate probe sets to clone Epo, and first actually attempting to do so, entitled him to the invention. On the particularly close question of obviousness, she rejected GI's argument that Lin was not entitled to a patent because the synthetic probe method was well known, and instead followed Ullrich's testimony favoring the Amgen position (and for which they could cite Maniatis himself as an authority): "although cloning seems straightforward on paper, it is more difficult to put into practice" because small variations in standard procedures can make all the difference.[36]

Thus, judged Saris, it was "obvious to try" screening with large synthetic oligonucleotide probe sets, as Fritsch and Biogen's scientists and others had all tried, but in 1983 no one had yet *successfully* cloned a gene from a genomic library using *two* sets of fully redundant synthetic probes as large as those used by Lin (128 different probes in each set). So there was no obvious assurance of success. As legal scholar Philippe Ducor has observed, the extreme care with which Saris—and subsequently the appeals court—treated the prior art of probing reflects the strenuous efforts of the courts to justify, within laws written to reward not just perspiration but inspiration too, patents for the practitioners of an increasingly routine industry of cloning natural genes to make protein drugs. (The problem has since faded in the realm of drugs, as genetic engineering firms have moved on from reproducing natural biochemicals to second and third generation proteins with new properties not found in nature. But it has recently come to a crisis in diagnostic tests for disease susceptibility using natural gene variants, because unlike natural protein drugs, for diagnostics these natural sequences cannot be improved upon. In the 2013 *Myriad* case the US Supreme Court, at the urging both of the executive branch and most life scientists, finally took steps to remedy that situation—not on grounds of obviousness but by ruling natural gene sequences per se to be unpatentable subject matter.)[37]

Saris's decision was symmetrical in supporting most of GI's patent on purified erythropoietin. Again, it was a close question of whether Hewick's innovation, using a marginally novel method of chromatography to purify Epo that had already been purified by the Miyake-Goldwasser method, was nonobvious. As with the Lin patent on the cloned Epo gene, Saris gave the benefit of the doubt to the patenting biotechnologist: while it was obvious in 1983 to apply RP-HPLC method to separate and purify glycoproteins, it was not obvious to apply the technique to Miyake's Epo "because . . . the scientific community believed that material to be pure already." Having pursued an uncompromising strategy, Amgen could now expect reprisals for all the various injunctions and restraining orders they had visited upon Genetics Institute. Or as GI attorney Bruce Eisen put it: "They set the ground rules—winner take all—and if they lose, they've got to be big boys about it."[38]

Sure enough, in early 1990 GI filed for an injunction to prevent Amgen from selling Epo in the US, since the protein infringed their patent. Two months later the Boston district court ordered Amgen and GI temporarily to cross-license their patents without royalty, and for Amgen to submit information to the FDA that would assist in GI's application to market Epo. Amgen appealed the order, a "delay tactic" (said legal observers) that worked to keep GI's Epo off the US market. While the crucial appeals and lawsuits continued, GI also petitioned the FDA to rescind Amgen's claim to Orphan Drug exclusivity for Epo, arguing that the firm's 1990 sales of $275 million—in its first full year on the market—represented an incentive far exceeding that intended by the underpinning legislation. Indeed, that sales already exceeded projections also suggested that many patients were receiving it off-label, that is, not for the FDA-approved indication of anemia caused by advanced kidney disease (even among the Medicare patients accounting for $200 million of those sales, at about $4,000 per patient).[39]

This dispute spilled over into Congress, then considering changes to the Orphan Drug Act, the 1983 law meant to encourage development of treatments for rare diseases that were therefore "of little or no commercial value." Initially, the law extended market exclusivity only to unpatented drugs with low sales, on the order of $5 million annually, but it was amended in 1984 and 1985 to extend protection even to patented drugs, and to treatments for conditions affecting fewer than 200,000 Americans independent of sales. To critics in 1990 it had become an unjustified subsidy to certain firms, protecting blockbuster products from competition—notably Genentech's growth hormone. Resuming a 1987 effort "to separate the true orphans from those drugs that have very wealthy parents" and sales in the hundreds of millions of dollars, originally blocked by the pharmaceutical lobby, consumer advocate (and original author of the law) Congressman Henry Waxman tried again with the backing of AIDS patient groups. His proposals would, among other amendments intended to restore the bill's intent, boost tax incentives but revoke Orphan market exclusivity when a condition's patient population exceeded 200,000 (as AIDS was then doing), and allow competing Orphan Drugs if they were developed nearly simultaneously. The biotechnology sector divided against itself. Amgen and Genentech predictably opposed losing their monopolies, Rathmann condemning the idea of breaking this "covenant with the pioneers." Genetics Institute and Cetus supported the changes and formed their own lobbying group of mostly smaller firms. They critiqued the act as abused by bigger firms and—as GI chief executive Gabriel Schmergel put it—basically flawed in its creation of a "winner-take-all" effect that allowed whoever happened to clone a gene first (by increasingly routine methods) to exclude rivals with less expensive or improved products, even if those newer products were patentably different. The struggle in Congress was intense, framed as a debate over whether the existing act harnessed market forces to save lives, or represented a market-distorting "license to profiteer." Epo, with its skyrocketing sales after only one year and its great cost to Medicare, was the prime exhibit. A watered-down amendment eventually passed both houses, but in late 1990 President George H. W. Bush (a former Lilly board member and a major recipient of campaign contributions from Genentech and Lilly) killed it with a pocket veto.[40]

Meanwhile the CAFC heard the appeals to Saris's decision that would, in the realm of patent law, effectively decide the same controversy about the shape of monopoly incentives for developing drugs through genetic engineering. In March 1991, Alan Lourie, a new Bush appointment to the Court, issued an opinion making key points about scope, conception, and obviousness in biotechnology patents. In his youth a chemist at Monsanto and Wyeth Pharmaceuticals, and later chief counsel for SmithKline Beecham, this judge certainly brought the wisdom of the established drug industry to bear. Lourie entirely upheld Saris's opinion on Amgen's Epo DNA patent, including both her extension of its scope to cover any DNA encoding the natural Epo protein (including the cDNA), and her limitation of its scope to exclude claims on unspecified DNA sequences encoding any protein whatsoever with Epo activity. Here he suggested that future patents could be sustained on Epo variants with improved properties, a hint Amgen took to heart in developing a second-generation Epo protein, featuring a slightly different amino acid sequence that altered the sugars on the protein to extend its action. (Brand-named Aranesp, this protein would be sold by Amgen in competition with its own first-generation Epo as marketed by Johnson & Johnson.) As for conception, Lourie made a particular point of upholding Saris's rejection of GI's argument that, because Fritsch could show he had earlier conceived of (and attempted) probing a genomic library with two large degenerate probe sets to clone Epo, the invention was his. Citing a GI expert witness, "you have to clone it first to get the sequence"; that is, the invention of the isolated Epo-encoding DNA occurred only when the probing idea was reduced to practice. Or as the implications of this reasoning were informally understood in the biotech world, "you haven't got it until you've cloned it." The important precedent set by this part of the decision makes common sense as policy: only the person who actually clones a gene to make a therapeutic protein should win patent rights, or else everyone would be discouraged from trying (because someone might later show that they had thought of doing the same thing at an earlier date).[41]

However, common sense or established jurisprudence cannot be so clearly perceived in the way Lourie's decision favored Amgen's isolated DNA patent over GI's purified protein patent. The CAFC could have specified a criterion for deciding which patents on isolated natural biochemicals should be valid, such as whether the preparation has some greater or different medical utility compared with those previously available. Such a criterion might have helped guide biotechnology research away from work aiming only to gain patent leverage on natural entities. Instead the court quibbled with the lower court's construal of an uncertain point in the Hewick purification patent, which had been treated by the lower court so as to give benefit of the doubt to the patent holder (as is standard). To be precise, Lourie decided that there was no convincing evidence that Hewick's patent showed how to purify Epo protein with a certain minimum specific activity measured in vivo, when the patent did not specify in vivo or in vitro.[42] In invalidating the Hewick patent on the same sort of minor detail that, without benefit of doubt, might equally have invalidated the Lin patent, the federal courts also missed the opportunity to introduce some competition into the already huge and lucrative US market for a drug derived from Goldwasser's taxpayer-funded work (in 1991 Amgen was already selling over $400 million of Epogen, returning profits at an astonishing rate around 30%). This "shocker" of a decision, astonishing both to legal and business analysts, sent GI stock into a 35% tumble, spelling doom.[43] Whatever the basic reason of Lourie's decision—prejudice favoring DNA over protein (common enough in the early days of recombinant enthusiasm), a genuine blunder in understanding factual evidence (as GI argued in an appeal petition),[44] or a desperate effort to remove ambiguity and thus signal that biotechnology offered blockbuster jackpots—it had the last effect.

In this and two subsequent, precedent-setting decisions, Lourie reinforced a third core message on the question of obviousness: scientists could win patent rights over essentially any natural protein that they succeeded in cloning first, no matter how well-known the cloning methods they used might be or how straightforward the path to success might seem. Here the CAFC was pushing back against the Patent Office, which in the 1990s was trying to raise the innovation bar on grounds of nonobviousness by denying patents on genes for previously characterized, naturally occurring proteins cloned by well-known standard methods. But Lourie and his colleagues effectively took obviousness off the table as grounds for challenging biotechnology patents (in line, to be sure, with language added in 1952 patent law by the pharmaceutical lobby). The CAFC's logic may have made a "fetish of unpredictability" as the antidote to obviousness, legal analysts observed, but it delivered on the court's perceived duty to reward those who "elect to invest the time and money" to prove a cloning strategy would work.[45]

Of course, the lower the bar on obviousness, so that relatively obvious and routine inventions can win patent monopolies of broad scope, the less the patent system encourages meaningful innovation. One constructive response to this problem might have been to emphasize innovation in methods and processes, where greater inventiveness would justify broader protection for products of the processes. But the CAFC's weak alternate answer can be found in its limits on the scope of patent protection for the proteins and their analogs that biotech firms actually produced, leaving out further variants not explicitly disclosed. As one counsel for a biotechnology firm described the implications of this package of decisions, "Judge Lourie seems to be creating a doctrine for 'gene jockey' cases which permits one to slice the salami thinly," allowing monopoly protection for almost every new biopharmaceutical meeting very low standards of novelty and nonobviousness—but at the same time offering the possibility of patents on very slightly improved products. This, of course, was the intellectual property regime that the big US pharmaceutical firms were accustomed to, one so often criticized for directing the bulk of research effort into trivially different me-too drugs.[46]

Conclusion: Law, Markets, and Publics

There was a time, in the 1980s, when the new technology of genetic engineering was expected to lift the pharmaceutical industry out of its me-too rut, boosting the flagging rate at which genuinely novel medicines were developed. This in turn would help rejuvenate American manufacturing industry, according both to enthusiasts in the business world and to more sober judgments from the Congressional Office of Technology Assessment. But as I have just argued, in the early 1990s the US Federal courts quickly assimilated the new biotechnology to the established intellectual property regime of the pharmaceuticals industry. This established regime, in richly rewarding trivially different new drugs, incentivized correspondingly "thinly sliced" innovation, and thus reduced incentives to academic-style research aimed at broad basic patents and especially method claims, such as that epitomized by the Stanford Cohen-Boyer cloning patents. In effect the American courts derived science policy from big pharma and imposed it on the small biotechs. To be sure, in the history of technology it may not be unusual for major decisions affecting the future of industries to be left to the patent courts, where they are especially liable to influence (perhaps even "interpretive capture") by the patent system's established users—the big companies and their lawyers.[47] Still, this outcome is a noteworthy sign of the times in the case of genetic engineering, a field that had been the subject of so much high-level policy discourse on how best to harness it for the greater national good. From the mid-1980s to the mid-1990s, for example, biotech had been the topic of ten OTA special reports. That in effect, entrenched corporations should make the policy decisions certainly fits with the tenor of policy in the neoliberal age. So too the OTA's abolition in 1995, depriving elected government of information on which to base intelligent policymaking.[48]

The CAFC's endorsement of a monopoly for Amgen on first generation erythropoietin, covering all products made using the natural Epo sequence, certainly established that the American legal system would sufficiently reward biotechnology drug development to attract further investment—perhaps much more than sufficiently, in this instance. Even at the time of Lourie's 1991 decision Epo sales were exploding and Amgen was reaping huge profit, and over 15 years the world market in Epo (first and second generation) reached $13 billion. Certainly no incentive so big required a second helping of government aid through Orphan

Drug exclusivity in order to attract research and development efforts. One need only consider the intense, expensive, and redundant Epo cloning efforts undertaken by Biogen as well as GI and Amgen by 1984, before the US Orphan Drug law was amended to extend protection irrespective of sales, and when the potential market was judged in the low hundred millions of dollars.

That the European Patent Office and European courts handled the claims on the same set of biotechnology innovations differently eliminates any sense of inevitably to the winner-take-all, ultra-high-stakes policy set by the American lawmakers and (especially) courts. Since European patent law is traditionally less generous to claimants on natural biological entities than US law, it comes as no surprise that GI's patent on the pure natural erythropoietin protein did not stand. Amgen's patent claims on the natural erythropoietin gene, the basis of its monopoly on first-generation Epo in the US, did not fall in Europe but their scope was more restricted to cover not all Epo-encoding DNA but only the human genomic DNA and monkey cDNA that the firm demonstrably cloned first. GI's patent claims on human cDNA also survived in the European Patent Office and many court jurisdictions, and the rights were acquired by Roche. And the Elanex Epo deriving from Powell's independent cloning effort survived as well, the rights to which were later bought by Baxter. The GI and Elanex drugs were given the generic names epoetin beta and epoetin omega, differing in mode of manufacture and slightly in glycosylation from Amgen's epoeitin alfa, but were clinically equivalent.[49] Thus, from the mid 1990s onward Europeans had greater access to different erythropoietins from competing makers (Amgen's product sold by Johnson & Johnson under the name Eprex; the product based on GI's patent, from its partner Boehringer Mannheim and then Roche, best known as NeoRecormon; and the product that Elanex named Epomax). And as one would expect in a more competitive market, prices paid by Europeans were lower.

Comparison with the European situation, with its narrower scope of patent protection permitting some competition in the Epo market, adds further weight to the view that the American intellectual property system has over-rewarded biotech drug development in at least this case—and to the point of producing major perverse consequences. Granted, other factors beside large monopoly profits are in play. The same national health insurance systems that negotiate reasonable drug prices in Europe also regulate drug advertising and guide prescribing more strictly. The net effect was that despite lower prices, Europeans consumed less Epo, particularly for anemias caused by some cancers and most cancer chemotherapy—the drugs' largest market. That is, anemic cancer patients in the

United States were less likely to receive transfusions than Europeans and more likely to get Epo, and when they received Epo got higher doses. Marketing effort drove this greater American consumption. For example, the US allows television advertising of prescription drugs to consumers, and it was used heavily by Johnson & Johnson to sell Epo for "fatigue" in cancer (rather than anemia) starting in 1999. Despite FDA reprimands for the unapproved efficacy claims in these ads, and Congressional concern that the advertising exploited the vulnerability of cancer patients, this marketing blitz was not curtailed for years—largely in deference to "commercial free speech." Similarly ineffectual were FDA reprimands to both Amgen and Johnson & Johnson for illegitimate marketing of unapproved high doses to doctors. Both firms also engaged in "rebate" schemes, which paid doctors as much as $1,200 per prescription (the cost of which was generally borne by Medicare or private insurers). Some medical practices collected millions of dollars worth of such "rebates" each year.[50] Thus, loosening American drug regulation provided the opportunity for aggressive Epo marketing, while the monopoly profits on the drug's sales provided both means and motive for the lavish campaigns. Between 2001 when Amgen's Aranesp first went on the market and 2006, annual worldwide sales of all Epo products doubled to a $13 billion peak (3/4 of which came from the United States, where the two Amgen Epo brands were the 4th and 6th best-selling prescription drugs). In that same interval medical usage (i.e., prescribing) increased only 51% in Europe as compared with 340% in the US.[51]

And European patients ultimately benefited from the restraints (lower profit incentive combined with stricter regulation) that limited their Epo consumption. In late 2003, worrying clinical trial data finally started to emerge. The drug did relieve perceived weakness in anemic cancer patients and reduce transfusions—the major trial outcomes on which approval of the drug in the US and Europe had been based. But erythropoietin was actually killing patients by stimulating many cancers. (Survival had not been reported in earlier, corporate-funded trials on Epo in cancer.) Soon afterward, it was also discovered that using high dosages of Epo to raise hemoglobin levels to near normal levels could cause an increase in thromboembolisms, that is, catastrophic clotting. Between the increased cancer and the thromboembolisms, cancer patients in the US, more often and more heavily treated with Epo as a supportive therapy than similar cancer patients in Europe, were dying at an approximately 10% higher rate than their European counterparts. Put another way, for the better part of a decade the overenthusiastic Epo treatment practices prevailing in the US shortened the lives of thousands of American cancer patients each year. Since then the use of the drug has been cut in half in the United States and cut slightly in Europe too, reducing the disparity.[52]

Thus the most important drug to yet come from recombinant DNA turned out to be a mixed blessing not only for the health systems that had to pay such high prices for it, but also for the sick people purportedly benefiting. That Epo came on the market so quickly and rose so high on the sales charts certainly owes much to the American intellectual property reward scheme for biotechnology since the 1980s: monopoly profits paid not just for the costly lab science and clinical trials needed to make Epo available, many times over, but also for the heavy marketing of the drug that pushed its consumption far beyond the patient populations most apt to benefit. Further, the delayed emergence of awareness about Epo's harms in the medical literature might also be counted among the social costs of the hyperenriched American reward system for drug development. Between 1988 and 2008 published lab studies funded by the manufacturers addressed the possibility that Epo contributed to cancer significantly more rarely than independent studies on the hormone's action. Like the extravagant marketing, this sort of effect on the medical research literature costs money too.[53]

tPA

The End of the Beginning

Since the retreat of infectious diseases in the early twentieth century, people in developed countries began living long enough to die of something different—especially cancer and heart disease. In the United States around 1980, about a million people each year suffered a visit from heart disease's main killer, that traumatic event popularly know as a heart attack and medically as a myocardial infarction. In an infarction the arteries feeding the heart clog with a blood clot, causing heart muscle to die, thus reducing the heart's pumping efficiency and depriving other tissues of oxygen. The victims were (and are still) mainly men middle-aged and older. More than one third died as a direct result of the event; for many others, it marked the end of life as they knew it. Of course, in the 1970s Americans had access to ever more complex and expensive life-extending technologies such as pacemakers, and heart transplants were then advancing from experimental trial to common practice. Total artificial hearts had entered clinical testing and seemed to be coming soon. But these same technologies that had once been hailed as medical wonders drew criticism in the 1970s for the high costs and low quality of life they often entailed. Indeed, for some the artificial heart was emblematic of everything that was wrong with medicine—a benefit to researchers, physicians, and medical manufacturers far more than to the patients whose lives they briefly extended. So cardiologists were looking especially hard for ways to prevent heart attacks. One of their main frontiers in the later 1970s was surgical prevention in those at high risk, with both bypass grafting and less invasive angioplasty procedures.[1]

But what if something, apart from giving the victim oxygen (and the temporary assist pumps then becoming common), could be done *during* a heart attack to limit its destructive effect on heart muscle? Using experimental animals, over the past decade cardiologists and hematologists had developed methods to restore blood supply during the first hours of an evolving heart attack by clearing the clot. Physicians were particularly excited about attacking the clot with a bacteria-derived protein called streptokinase, which activates the body's own clot-dissolving system. To reduce the risks of hemorrhagic strokes and other bleeding throughout the body, the drug was released near the coronary clot, using a catheter tube snaked through blood vessels into the heart. This catheterisation procedure required high technology, specialised skills, and a good deal of time. So, although by 1981 thrombolytic (clot-breaking) therapy had been shown through pilot studies in humans to work, it seemed unlikely ever to become a routine procedure for most heart attacks, where, after all, intervention was likely to come in an ambulance or in a simple emergency room, neither of which would be equipped to guide a catheter swiftly into the right part of a critically ill patient's heart. Even so, to some this "intracoronary thrombolysis" technique seemed the "wave of the future"—if only it could be simplified, and ideally made into an easily administered drug.[2]

Thus cardiologists and other physicians wanted a thrombolytic drug for heart attacks, and perceived that one might be within reach. But they could only obtain it through the lab science of drug firms and also that strange *pas de trois*, encountered in chapter 4, of the commercially sponsored clinical trial. Here government drug regulators require tests proving scientifically that a new drug benefits patients as much as existing therapies, before allowing a manufacturer to sell the drug for a certain condition. The company must supply the drugs, and engage, and usually also fund physician-researchers (preferably influential ones already inclined to believe that the drug will work) in assembling a trial that can show their product to be effective. These physicians then conduct the study, using the drug on actual patients under their care according to the design of the trial. What is odd about this dance is that the wishes of each participant often clash, and none plays the unambiguous lead. The regulatory agency, which is supposed to serve patient health according to medical science and therefore cannot contradict expert medical opinion, does not have the power to tell such physicians what tests they should do for what illnesses. For their part, medical experts cannot force a drug firm to let them try the experimental, commercially unavailable medicine—let alone make it pay for a study designed to answer *their* research questions ideally. At the same time, drug firms that disregard expert medical opinion cannot expect either regulators, who rely on expert assessments, or ordinary physicians always to accept the outcomes of trials designed only to meet their marketing needs. Therefore, controlled clinical trials—especially the Phase III trials still officially required for initial marketing approval—typically represent a negotiated compromise between the sponsoring firm's business goals, the politically imposed (and therefore malleable) standards of regulatory agencies, and what at least some influential medical experts in a field desire.[3] For biotech firms bringing their first drugs to market, this complex, expensive dance was unfamiliar, and without the revenue streams of established drug firms they were poorly equipped for its protracted timetable. Many stumbled, and for those that did not the price of success would be unexpectedly high.

With a million possible consumers each year, and their cardiologists enthusiastic about thrombolytics, pharmaceutical firms were alert to the potential. Initially approved for dissolving clots elsewhere in the body, several thrombolytic products were already on the market by 1980. These were quickly slated for clinical testing for heart attack by their makers, to establish whether an effective dosing regime could be found that avoided both catheter delivery and the anticipated problem of generalized bleeding. Along with streptokinase, a 30-year-old product sold in the United States by tiny Kabi and giant Hoechst, there was also the human enzyme urokinase. The two firms that sold urokinase in the US, Abbott and Sterling, offered products manufactured by different methods. Abbott produced it with cultured human kidney cells, whereas Sterling imported it from a Japanese firm that extracted it from urine via the same type of industrial-scale hospital collection and purification systems used to harvest Epo in Japan.[4]

Less well known was another blood protein called tissue-type plasminogen activator or tPA. Discovered by Danish physiologist Tage Astrup and his colleagues in the 1940s, the substance was an important building block of the concept that the blood contained a natural system of checks and balances to maintain the optimal level of clotting, but unlike streptokinase it had never been available for clinical testing. Still, to hematologists and others aware of this laboratory curiosity in the 1970s, tPA was expected to be superior as a thrombolytic drug: whereas urokinase and streptokinase generated the body's own clot-dissolving enzyme plasmin (by converting it from its less active precursor, plasminogen) throughout the bloodstream, on the basis of lab evidence tPA appeared to exhibit this activity only in the presence of the major protein in clots, fibrin. This suggested that tPA might trigger thrombolytic action only where clots already existed. Also, it seemed that tPA did not attack fibrinogen, the circulating precursor protein to fibrin, and therefore did not reduce clotting overall. This specificity together with its key clot-seeking feature implied in turn that it might be safe simply to inject tPA into the bloodstream without worrying much about the dosage, making it a real possibility for first-line treatment of heart attacks even by paramedics.[5]

Streptokinase was cheap to make from bacteria, but because of high production costs, drug firms soon began to look at genetic engineering as a way to make urokinase. By 1980 a cloning race was underway, but with a major difference from the previous cloning efforts: this time, established pharmaceutical firms directly entered the fray. Kabi, Genentech's European partner in growth hormone and already a vendor of urokinase, launched into cloning tPA on its own around 1980. Abbott attempted to clone human urokinase. And the German company Gruenenthal contracted with Genentech to clone the urokinase gene too, on terms resembling Genentech's growth hormone contract with Kabi.[6] The race to bring these expected medical and commercial breakthroughs to market would in a sense launch the next phase of biotech, in that it spurred research into second-generation protein drugs engineered with altered properties, while it also less predictably altered the nature of the biotech firms.

Chasing the Billion-Dollar Protein

In 1980 Herb Heyneker, now a senior scientist at Genenech, took up the urokinase cloning project. The mRNA was rare, and the effort to find it among cDNA clones went slowly. To help with this project, he hired Diane Pennica, fresh from nearly two years as a postdoctoral fellow in what she experienced as the academic-style, basic research environment at Roche Institute in New Jersey. Her project there had been to isolate and study transcripts of an animal virus, and she proved adept at the isolation of delicate mRNA with electrophoresis gels. Within a few weeks of Pennica's arrival at Genentech, Heyneker asked her if she could attend a June meeting on clotting and thrombolysis in Sweden, in his place. Her charge was vague: to learn what was going on in the field, to see if there were promising drug candidates beyond urokinase, and if so, to see if anyone at the conference had the key ingredients for product development: a pure protein, a specific antibody or other straightforward way of detecting it, and most of all a cell line that overproduced the message to facilitate cloning.

Upon her arrival in Malmo, the confused Pennica accidentally walked into an invitation-only preconference seminar. There she found everything Genentech sought in a presentation by Belgian medical researcher Désiré Collen, who had just made great strides in producing and purifying tPA protein, both from human uterus tissue and (more tractably) from cultured cancer cells. She quickly convinced Collen to work with Genentech in cloning tPA from these cells, the widely available Bowes melanoma line, and began the project with Heyneker. Soon after, probably late in 1980, David Goeddel asked if she could take time off from tPA to help him make cDNA for his gamma interferon cloning project. And thus, as Pennica put it, she "left Herb's lab" and switched to Goeddel's, soon afterwards taking with her the tPA project (while Heyneker continued with urokinase). Such a way of framing this move, exactly as any postdoc in a university setting would put it, conveys how Pennica experienced Genentech as an informal academia-like institution—in these early days still a scientists' republic.[7]

The challenge with tPA was the protein's size, at more than 500 amino acids three times as large as growth hormone. That meant that its mRNA would be long as well, and proportionately more difficult to capture intact. And it was rare too, even in the Bowes melanoma cells that overproduced it—less than one in a thousand transcripts, almost as rare as the elusive Epo message. But in other ways tPA was not so difficult. There was a good antibody to help identify the protein, removing the need for a complex bioassay like many interferon cloners were using at the time. Pennica, Goeddel, and colleagues were among the first cloners to take up the synthetic probe set technique to screen their libraries, but had an easier time obtaining pure protein to sequence from the Bowes cells than the GI and Amgen groups cloning Epo one to two years later, and a much easier time finding useful low-degeneracy segments of amino acids in their protein fragments (ultimately succeeding with a single pool of eight 14-mers). To enrich for tPA message in cloning, Pennica identified the correct mRNA band by separating the various messages by size with a urea gel, extracting the most likely messages (by size), translating them in vitro, and testing them against the tPA antibody. Complementary DNA made from the tPA band was then spliced into pBR322 by G-C tailing.[8]

Even with these advantages it took Pennica a year of seven-day work weeks, following Goeddel's example, to find a clone in late 1981. Although a massive 2,300 bases long, containing a 1,500-base coding region as well as a generous 3' tail, the cloned cDNA lacked the 5' region that specified the first part of the tPA protein. Recovering that remainder of the tPA message took another half year via a circuitous route: Pennica and Goeddel made a synthetic 16-mer complementary to mRNA sequence at the leading (5') end of the clone they did have, hybridized that to the Bowes messenger RNA, and used reverse transcriptase to produce new cDNA reading from this synthetic primer into the unknown end of the message. This they cloned into pBR322 as before, now screening the resulting clones with a tPA genomic clone they had recovered from the Maniatis phage library (using a fragment of their previously cloned partial cDNA as a probe). Around April 1982 they recovered a small cDNA clone that included the translation start codon in the tPA leader sequence, 35 amino acids upstream of the beginning of the known tPA protein sequence (and contiguous with it). Much as they had with growth hormone and interferon alpha, the Genentech lab then pieced together a full length, functional tPA gene from the two cDNA clones, deleting the leader sequence and using the patent-pending synthetic DNA method to reconstitute the codons for the first few amino acids of the mature protein, plus an ATG start codon. They then tacked the whole 528-codon monster onto a high-expressing *trp* promoter. In *E. coli* this plasmid produced tPA—but not medically usable tPA, since it lacked the sugars carried by the natural protein. The yields, too, were poor, as once again *E. coli* packed the huge, alien protein into "refractile bodies." Still, the cloned sequence (including several errors) and the barely functional *E. coli* expression vector they had in hand were hurried into a patent application in early May 1982.[9]

Pennica first presented the results soon afterwards in July 1982, in Lausanne at the conference of the same society where, two years earlier, she had met Collen. On one slide she showed a probable structure for the full tPA protein, as predicted from the amino acid sequence indicated by her cloned DNA (but revealing only about 50 amino acids—less than 10 percent of the sequence—to protect Genentech's intellectual property). The slide show brought her a standing ovation. It was a major scientific accomplishment, even if not completely disclosed to the scientific community. At that point the firm issued a press release announcing their success, but CEO Robert Swanson barred Pennica from publishing its details, feeling that there was heavy competition and too much at stake. He was entranced with the idea that this drug, which he thought of as "Drano [drain opener] for heart attacks," would (at about $2,000 per patient) be the billion-dollar blockbuster that Genentech would market for itself and ride to the pharmaceutical major leagues.[10]

For the rest of 1982, Pennica and colleagues worked on transferring her full-length semisynthetic tPA gene into yeast, or into animal cells that would add the necessary sugars, and boosting expression to useful levels. But in the autumn, an event occurred that says much about the extent to which Genentech then enjoyed status as a honorary academic institution: a biologist telephoned Pennica because a journal had sent him a paper for review describing the cloning of tPA. Haven't you cloned this already, he wanted to know? If so, and there was a paper in press reporting it, he was willing to reject the manuscript—a courtesy that academic biologists might typically extend one another. Pennica

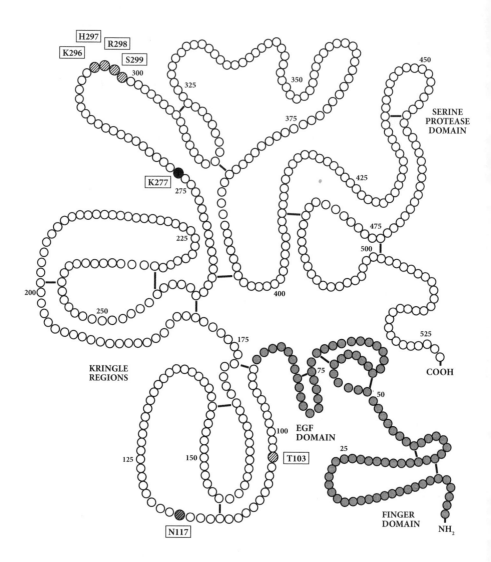

admitted that she had not yet submitted a paper. After she "yelled" at Swanson for allowing her to be "scooped," another sign of the informal academic atmosphere then prevailing at Genentech, he authorized her to submit a report to *Nature* post haste. Perhaps through benign editorial conspiracy, the competing paper appeared in *PNAS* in January 1983, simultaneously with the Genentech *Nature* paper.[11]

The *PNAS* competitors were a team from Swedish universities and Kabigen, the new biotech subsidiary of Genentech's erstwhile collaborator Kabi, who had used the same Bowes melanoma cell line to obtain both message and tPA protein for sequencing. From an internal six-amino-acid sequence with low redundancy, they had created two small, fully degenerate oligonucleotide probe sets and recovered one 370-base fragment of the tPA gene by screening several thousand bacteria transformed with pBR322 carrying Bowes cDNA. This was basically the same cloning strategy that Genentech used, except that Kabigen researchers had not enriched for tPA message to the same extent. But they had only isolated a partial clone, some 15% of the message; by contrast, in their *Nature* paper the Genentech group reported cloning the whole thing, and sequencing it, and its predicted protein structure (fig. 6.1)—and achieving a little expression too. Soon after the competing publications, the Kabigen group personally visited Pennica in San Francisco, conceding defeat graciously. In April 1983 Genentech amended Pennica's patent application from the year before, correcting some sequence errors and additionally claiming expression of their cloned tPA in Chinese hamster ovary (CHO) cells, using a vector carrying mouse DHFR and simple SV40 control elements (similar to Amgen's Epo expression system; see fig. 5.1). Time was of the essence.[12]

Fig. 6.1. The structure of the natural tissue plasminogen activator (tPA) protein as predicted from the full length cDNA clones obtained in 1982 by Diane Pennica and David Goeddel, and commercialized by Genentech under the brand name Activase. Boxes indicate changes introduced to produce second-generation tPA products as follows: in 1984, Joseph and Margaret Rosa at Biogen substituted isoleucine for the lysine at position 277, reducing autoactivation; around 1986 Glenn Larsen and a team at Genetics Institute substituted glutamine for asparagine at position 117, eliminating the glycosylation site, and deleted the growth factor and finger domains before position 92, increasing fibrin specificity (though reducing absolute affinity); and around 1992 a team at Genentech created a variant tPA containing the same glutamine-asparagine substitution at position 117, a substitution of asparagine for the threonine at position 103, thus introducing a new glycosylation site, and the substitution of alanine for all four amino acids at positions 296–299, increasing fibrin specificity. (See US patents 4,753,879, 5,071,972, and 5,612,029, *Heart* 2003;89:1358–62, and main text for further details.)

As Genentech scientists well knew, they had other competitors. At the July 1982 Lausanne meeting, a photographer set up a tripod in the aisles and photographed Pennica's slides during her presentation. This unsporting if ingenuous act was the doing of Robert Kay, whose team at Genetics Institute had been trying to clone the protein for six months. Two years earlier Kay had been a postdoc at the Imperial Cancer Research Fund in London, where he had cloned a hemoglobin cDNA from tadpoles, determined the full sequence, and shown that this protein differed significantly from adult frog hemoglobin. This was important, basic molecular biology of the day, focussed on the developmental regulation of gene expression. It also showed technical mastery of eukaryotic cDNA cloning, bringing him unsolicited job offers from Amgen, Genetics Institute, and a Stanford-based firm called DNAX that Paul Berg and Charles Yanofsky were founding, all at the start of 1981. Signing with GI, Kay there carried forward a tPA cloning project that the company had picked up from a short-lived biotech firm named "Cellbiology," based at Cold Spring Harbor Laboratory and founded by none other than James Watson and his colleague Joseph Sambrook. In mid-1980, Watson and Sambrook had decided to pursue tPA; by the beginning of 1982, with funding from the blood products firm Baxter-Travenol, Sambrook and collaborator Mary Jane Gething (then visiting CSHL from Britain) were screening Bowes cDNA spliced into expression vectors, using antibodies to identify any *E. coli* clones producing tPA protein. At about this point, near the start of 1982, Genetics Institute joined the Baxter-funded tPA collaboration and took on the alternative cloning strategy of screening with synthetic probe sets.[13]

The GI team, like all their competitors, was screening Bowes melanoma cDNA spliced into pBR322. Protein chemist Rodney Hewick derived the amino acid sequence of purified tPA from Bowes cell cultures for the design of degenerate probe sets, the same technique that Fritsch would use with the Epo project that GI also initiated that year. Glenn Larsen, a freshly minted PhD who led the actual cloning in Kay's group, used these to screen both his own transformants and some made earlier by the CSHL group. On his return from Lausanne in August, Kay was disappointed with the resolution of his photographs of Pennica's slides, but Larsen greeted him with the news that they finally had a clone. This clone, retrieved with a probe set covering six amino acids near the leading end of the mature tPA protein, was incomplete. But by using an internal portion of that tPA clone as a probe they were able to obtain a second clone, and from that at a third, so that they were able to splice together a complete tPA cDNA and show expression of tPA protein with it by January 1983, at just the same time that Pennica's and Kabi's reports were published.[14]

With their final cDNA assembled and its sequence confirmed in an expression vector that May, GI immediately moved on to the next lap of the race: expressing tPA efficiently, preferably by integrating it stably into the chromosomes of a mammalian cell line that would add the correct sugars. In the second half of 1983, GI scientists including Randal Kaufman, whose PhD work at Stanford with Bob Schimke and just-finished postdoc with Phil Sharp at MIT had made him an ace engineer of animal gene expression, moved the natural tPA cDNA into a vector to express and amplify the gene in CHO and other mammal cells. This expression system was more sophisticated than those developed around the same time for erythropoietin at Amgen and also at GI (fig. 5.2). Indeed, combining elements such as an immunoglobulin and adenovirus enhancers to boost expression, together with replication and transcriptional control sequences from SV40, Kaufman's tPA expression system was a scientific accomplishment in its own right.[15]

Impressed, the large British firm Wellcome (then a world leader in large-scale mammalian cell culture) in 1984 bought Baxter-Travenol's interest in GI's cloned tPA, even though the company lagged about six months behind Genentech. In 1985 Kaufman, Larsen, and other GI scientists worked closely with Wellcome to transfer the CHO cell lines and get them producing tPA in Britain. By the first half of 1986 Wellcome and Genetics Institute had formed a joint venture to manufacture their first-generation tPA, won FDA clearance on their manufactured tPA, and begun Phase I testing of the product's human pharmacology. At Genentech, the scientists were worried; GI was the "competitor that mattered" (Kleid recalled) and they were catching up.[16]

But GI was not Genentech's only major competitor. Events at Biogen, too, had followed a similar trajectory. Using the same methods as their competitors, by 1983 a Biogen team in the US had independently cloned the natural gene and expressed it in a bacterial plasmid, at first with a short leader. Biogen's tPA project was funded by the Japanese firm Fujisawa, in exchange for rights to market the drug in Asia. Once the correct natural sequence was cloned and expressed, in 1984 Biogen sold the remaining rights to the large American drug firm Smith, Kline and French (soon SmithKline) in exchange for a funding infusion as well as eventual royalties. While further development of the natural tPA gene went ahead elsewhere in Biogen and SmithKline, Biogen's protein chemist Joe Rosa—one of the firm's first hires (from a Yale postdoc) to its Massachusetts operation—moved on to the third lap in late 1983, pursuing a second generation tPA project.

Rosa's project is noteworthy as an early example of "rational drug design," with ideas from physical biochemistry driving the cloning work. Rosa observed that, like other members of the serine protease family of enzymes, tPA was activated by cleavage of its amino acid chain at a particular point in its sequence. In serine proteases the activating cleavage typically freed a positively charged amino acid near the enzyme's active site, but tPA was unusual in that it had some activity even before being cleaved. By modelling the three-dimensional protein structure, Rosa surmised that a lysine amino acid (which is positively charged) at position 277 might be responsible. If so, eliminating it might reduce the protein's biological activity in general blood circulation, thus improving the protein's specificity for clots. He and his partner, molecular geneticist Margaret Rosa, created mutant forms of the natural tPA by replacing lysine with neutral amino acids at position 277. For this they used the new method of site-directed mutagenesis, where a synthetic DNA strand with a slightly different desired sequence is introduced, such that the bacterium replaces the original through its own repair mechanisms. As intended, the second-generation tPA developed in 1984 had a longer duration of action in the blood, and seemed to activate more specifically at clots. Its pharmacological advantages, however, were not enough to tempt the firm to develop it further in clinical testing. Meanwhile, Smith-Kline and Biogen pushed their first generation recombinant tPA into clinical trials without delay.[17]

Battle for the Heart

With so many competing recombinant tPAs in the pipeline and no patents yet issued, the winner might effectively be decided by who got a drug to market first. By 1985, Genentech's tPA was close to commercial introduction. Collen, in keeping with his long investigation and advocacy of tPA for thrombolytic therapy, had carried his collaboration with the firm into preclinical (test tube and animal) testing of Genentech's recombinant protein once it became available from CHO cells in late 1983. For example, he showed that the Genentech tPA behaved the same as natural tPA from Bowes cells with respect to its fibrin binding properties. Collen also showed that infused recombinant tPA cleared coronary clots in dogs better than urokinase did, and with much less effect on circulating fibrinogen—and therefore less risk of promoting bleeding elsewhere. Such findings were most promising.

At the beginning of 1984 Genentech had received FDA approval to begin clinical testing of their tPA, and Collen participated in the first comparative Phase II trial using the drug with people suffering heart attacks. In this study at three US medical centers, 45 patients were randomly given either an intravenous infusion of tPA or a saline placebo, each lasting 30 to 90 minutes. Three-quarters of the patients treated with tPA experienced clot clearance, as compared with none of the untreated patients. These results were not uniformly positive: in many patients the clot returned after their tPA infusion stopped, and circulating fibrinogen was reduced too, on average, by 8%—not a negligible amount. The trial suggested limitations to Swanson's vision of tPA as a billion-dollar wonder drug: the slow infusion process implied that tPA's delivery would still require specially trained professionals, while the mounting evidence that tPA degraded fibrinogen implied that high doses would promote bleeding. But these were preliminary results, and they were acclaimed when first presented at the American Heart Association conference in November 1984.[18]

There was obviously considerable room for improvement in the first-generation product. From a commercial perspective, the most important factor would be the ability to deliver tPA as a simple, single-injection "bolus." Compared to other firms then investigating second-generation tPA products to achieve this aim, the project Larsen undertook at Genetics Institute was an especially bold and extensive effort to reengineer the protein so as to alter its pharmacological profile. TPA consists of five distinct domains stitched together, each one a type found in many proteins and thought (as Gilbert had first proposed in 1978) to have been assembled by evolution in modular fashion. In addition to the serine protease domain, which cleaves the target protein (plasminogen), tPA includes a domain called a finger, or F (for its shape), a domain called EGF or E (after the protein where it was first characterized), and two domains called kringles (for their twisted pastry-like configuration) (fig. 6.1). At an early stage in his analysis of tPA, Larsen used site-directed mutagenesis to delete either the F domain, the E domain, or both, then expressed and purified the mutant proteins. He found that the mutant tPAs behaved much like natural tPA in the test tube, opening the way to changing these domains without losing the protein's clinically valuable preference for fibrin over fibrinogen. Larsen then made 35 different mutants with small changes to the F domain, trying to find variants that would bind fibrin well, but persist longer in the bloodstream. The minor deletion mutants were not especially superior to the mutant lacking both F and E domains, which showed a higher relative binding preference for fibrin over fibrinogen (despite reduced absolute affinity). In another study, he altered the sites where sugars were added, again in an attempt to lengthen the amount of time the protein stayed active in the bloodstream. Two double domain-deletion forms with reduced glycosylation (named deltaFE1X and deltaFE3X) persisted much longer and were selected for further development.[19]

The therapeutic advantage implied by these altered properties excited Désiré Collen who, despite his financial interest in and reputational association with Genentech's first generation tPA, became a collaborator with GI and a major supporter of its second generation plasminogen activator project. Collen helped demonstrate the deltaFE1X and deltaFE3X mutants' longer duration of action, and showed that their finger-independent binding took place via one of the kringle domains (whereas natural tPA binds fibrin with both finger and kringle domains, with different dynamics). At some point in 1988 deltaFE1X was chosen as the best drug candidate because of its longer persistence in the bloodstream, which would allow delivery by a single bolus injection of a dose that, in theory, would clear heart clots with less generalized bleeding than tPA. GI renamed it nPA, for novel plasminogen activator, and in late 1988 GI announced they were bringing Larsen's creation into clinical trials in Japan with funding only from their Asian backer Suntory. They were still looking for a partner to bring it to the rest of the world. Its future in medicine looked bright—except lawsuits blocked the way.[20]

Clotting in the Courtroom

While still impatiently awaiting both FDA approval to market its anticipated blockbuster and its US patent, in early 1986 Genentech was issued a broad British patent with claims covering pure human tPA protein, any tPA produced by recombinant DNA, and any vector containing a tPA-coding sequence. Immediately, Wellcome sued Genentech challenging the patent and Genentech countersued Wellcome and GI for infringement. In June and July of 1987 the case was argued before the UK's High Court, with Watson in the audience and the whole world of molecular biology looking on. Controversy mainly hinged on how novel and inventive Pennica's cloning of tPA was. Maniatis supported the Wellcome line of argument, that using small degenerate synthetic probe sets (and everything else Pennica and Goeddel did) was obvious and commonplace in 1981. Genentech scored a coup by recruiting Paul Berg to contend essentially the opposite—or rather, that the combination of techniques was new and not obvious, even if each one was not. The Court ruled that the majority of Genentech's claims were, on grounds of obviousness, "plainly invalid" because they were too broad. As British journalists explained, the Court found Genentech "greedy and over-reaching." Clarifying that the core issue was the scope of protection merited by the inventiveness involved in cloning a natural sequence, the judge explained that "had they produced some new and valuable variant of tPA, they might have got protection on that;" however Genentech had only discovered "a particular route to a known end"—the natural gene.[21]

So it seemed that while first-generation recombinant drugs would receive scant patent protection in the UK, the High Court signalled that protection would be extended to genuinely improved second-generation drugs. Eventually, as we have seen in the US appeals court's treatment of the Epo dispute, the US courts offered more generous patent protection for first generation products, as was widely anticipated. But in 1987, no one knew how the US courts would treat patent claims to second-generation variations upon natural genes and proteins that were not specified in patents on the natural gene. Both investors and competitors seem to have sensed that the legal winds were favoring second-generation products, Genetics Institute and Integrated Genetics using the occasion of the British decision to announce their second-generation tPA candidates (Integrated Genetics having used Pennica's published sequence information in their cloning, just as they used Fritsch's published sequence in cloning Epo). But for the moment Genentech still held the lead in tPA, and defended its position aggressively so as to slow, if not exclude, the many competitors "nipping at its heels."[22]

Genentech appealed the High Court decision, rather than narrow and resubmit its UK patent application, so as to be maximally "troublesome to our friends at Wellcome" (as Genentech chief counsel Kiley put it), losing again in mid-1988. Meanwhile the firm launched tPA on the US market under the brand name Activase, with a massive marketing push in late 1987. In June 1988, on the same day that the Patent Office issued a US patent to Collen on the purified tPA protein, Genentech (which had licensed Collen's patent) filed suit against Wellcome and Genetics Institute. Genentech's patent on Pennica's cloned tPA was also granted shortly thereafter. At stake in the ensuing infringement litigation, much as in the Epo cases already underway, was the breadth of patents on natural genes in the United States. Perhaps alarmed by the broad scope of the US tPA patents issued to Genentech, SmithKline quickly terminated not only its collaboration with Biogen to develop their first-generation tPA, but also another codevelopment contract for a second-generation tPA (with Damon biotech). In 1990 the jury hearing the district court case in Delaware justified such reactions by handing a major victory to Genentech, deciding that both tPA products infringed, even though neither GI's second generation nPA nor even the first generation tPA cloned by GI and manufactured by Wellcome (which had one accidental amino acid substitution) were literally identical to Genentech's patented natural-sequence tPA. Wellcome dropped its collaboration with GI, which had just been extended to include nPA. In Asia, however, Suntory continued its partnership, despite legal challenges there. In the rest of the world, Genentech had effectively frozen development of second-generation tPA products through lawsuits.[23]

Both GI and Wellcome appealed the Delaware decision in 1992, perhaps stimulated by the Court of Appeals for the Federal Circuit's (CAFC's) "salami slicing" encouragement of second-generation products in its EPO decision of 1991 (chap. 5) and also by Genentech's admission in a lawsuit against another firm that its first-generation tPA was substantially different from Wellcome's. This case brought the issue of novelty and scope of patents on natural genes and gene products before a three-judge panel of the CAFC that included judge Alan Lourie, this time with a focus on the "doctrine of equivalents" (the grounds for the Delaware decision). In 1994 the CAFC panel remained true to Lourie's Epo precedent and overturned the Delaware court's broad interpretation of Genentech's claims. The ruling held that since Larsen's nPA was significantly different both in amino acid sequence and in biological properties to tPA, this second-generation tPA did not infringe on Genentech's patents. Underscoring the signal he wanted to send to the industry, Lourie added a special note in concurrence with the panel's opinion: when a patent on a natural gene specifically discloses only the sequence of that natural gene, its scope cannot reach beyond that sequence and "insubstantial" variants. In contrast, Larsen's nPA, with an amino acid sequence that differed by 15%, a different structure, and different pharmacological profile—and costing $20 million to develop—was hardly an insubstantial variant.[24]

Thus did the CAFC confirm its doctrine for "gene jockey cases" appealed from lower US courts in the early 1990s (chap. 5). First comers cloning natural genes would receive strong protection on resulting first-generation drugs, but would not be allowed, through either method or product patents, to block development of second-generation proteins with different properties. There must be room for the sort of competition familiar to the established drug industry in which, for better or worse, minor variants have long constituted most of its new products. In effect, so long as these slightly different products were good enough to pass muster with the FDA, the medical marketplace should decide their value. Having retarded competition from advancing tPA analogs as long as possible, at the end of 1994 Genentech commenced Phase I clinical testing of its own second-generation single-bolus product (dubbed tenecteplase), ten years after Activase and six years after Genetics Institute's nPA. Undeterred by Genentech's saber-rattling about its own pending patents on tPA analogs, in 1995 Bristol-

Myers Squibb (BMS) bought Wellcome's rights to GI's languishing nPA (now dubbed lanoteplase) and moved it into Phase II human testing, while Suntory commenced Phase III controlled trials in Japan. But faced with competition from Genentech's second-generation drug, lanoteplase was not moved closer to market after BMS's first Phase III trial in Europe produced, at great expense, a disappointing and ambiguous result. Some in GI felt the dose employed had been needlessly high, and that with a slightly different protocol the drug's benefits would have outweighed the extra bleeding it caused. That is, lanoteplase had crashed on the same obstacle that, if not for skilful trial design by Genentech's new pharma-seasoned management, nearly wrecked Activase (as we will see).[25]

Making Medicine and Markets

While the major pharmaceutical firms had, for half a century, developed elaborate and powerful ways both to monitor and influence the prescribing behavior of doctors, this sway was beyond the reach of small biotech firms. Their scientist-founders and eminent advisors in the early days did have the social authority and leverage to influence Congress, the courts, and the investment market, as we have seen. But when it came to medical practice, the biotech firms had to ride on the coattails of clinical thought leaders and impresarios, many of whom worked closely with major drug firms. In the case of interferon, the large firms with whom the small biotechs had partnered did the complex, conjoined tasks of running clinical trials, winning regulatory approval, and marketing the new drugs to physicians. Perhaps Genentech adventured beyond existing medical need, as perceived by most doctors, in marketing growth hormone through sponsorship of trials for conditions other than pituitary short stature (chap. 3). But there the medical speciality, paediatric endocrinology, was tiny. With thousands of cardiologists and a million heart attacks per year in the United States, Genentech and its tPA competitors had to influence medicine on a much larger scale if they wanted to succeed. This effort to play in the pharmaceutical big leagues would prove extremely treacherous, and ultimately fatal to Genentech in the form it then existed.[26]

The most important of Genentech's preapproval trials for Activase involved a close partnership with prominent cardiologists. As noted above, cardiologists greeted the results from Genentech's earliest Phase II trial, designed mainly to determine the best dosage regimes to obtain clot clearance, with enthusiasm. Shortly thereafter Elliott Grossbard, a former Roche executive who had been hired in 1982 to lead Genentech's tPA development, negotiated with the leaders of the Thrombolysis in Myocardial Infarction (TIMI) study group—an especially influential group of clinical investigators lead by Harvard cardiologist Eugene Braunwald—to include Activase in a trial designed to assess the effects of streptokinase or urokinase compared with placebo. Based on the lab work of Collen and others, the TIMI investigators expected the results from tPA to be superior to those of urokinase, so Grossbard mainly had to convince them to wait until his firm's recombinant product was approved for testing by FDA and available in sufficient quantities. Redesigned to fit both Genentech's and the cardiologists' interests, the TIMI-1 trial became a comparison of streptokinase to Activase, split into a typical exploratory Phase II trial to help Genentech find the optimal dosing regime, and then a larger double-blind comparative trial (in current terminology, a Phase II/III trial, with a randomized design capable in principle of proving efficacy, but with small patient numbers).[27]

Counting on the convergence of the cardiologists' interests with his own, Grossbard had consciously taken a gamble in working with the TIMI group. Because of the study's prominence and its NIH affiliation, Genentech could not expect to optimize the trial's design to its product's best advantage. Nor could it expect to contain unfavorable outcomes should they occur. The gamble paid off in February 1985 when the 300-patient Phase II/III trial was stopped early because patients receiving tPA over 3 hours showed twice streptokinase's rate of coronary artery "patency" (or reopening, as determined by angiography)—the trial's main outcome measure—and no cases of hemorrhagic stroke. Effects on heart function and patient survival after treatment were not assessed. The TIMI investigators were as pleased as Grossbard, and they rushed a special report into print in the *New England Journal of Medicine* that sent Genentech's stock up 50%. Kirk Raab, Genentech's new president, built a massive new factory to produce tPA and stoked stock market expectations of a blockbuster.[28]

Notwithstanding its implied seal of NIH approval, and however impressive an advantage it demonstrated for tPA in terms of quick artery-opening, the TIMI trial was not quite enough for FDA marketing approval despite the loosening regulatory stringency of the time. In 1985, US regulations—at least outside of the politically supercharged areas of cancer and AIDS—required two well-controlled studies demonstrating efficacy in clinical terms, that is, improvement of the patient's condition. While nobody disputed TIMI's objectivity, the outcome it measured (coronary patency) was of debatable clinical significance. Perhaps patients had another heart attack soon afterward, or suffered delayed effects from the first attack, leaving them in the same condition whether or not they had been treated with tPA.

Genentech decided that it did not want to address this doubt about tPA, pinning its bid for approval on the patency data alone. This choice was strictly a business decision and had nothing to do with expert disinterest. In response to a large Italian trial showing improved survival with streptokinase compared with placebo, the TIMI group pleaded with Genentech to supply them with tPA for a similar, larger Phase III trial to establish whether the drug was clinically superior to streptokinase in terms of mortality. The firm refused, on reasoning that with American cardiologists already convinced of tPA's superiority on the basis of patency, the FDA would follow medical enthusiasm and approve the drug; thus Genentech had nothing to gain and much to lose by assessing whether their product improved long-term patient survival. As Grossbard recalled, in an interview with journalist Linda Marsa some ten years later, when in late 1985 TIMI representatives suggested that Genentech sponsor a mortality-assessing trial he replied: "I think that's a terrible idea. I'll present it to Kirk [Raab] and Jim [Gower, Genentech VP]. But I'm going to tell them not to do it. Thanks to you almost every cardiologist in America is convinced that tPA is so good that you had to stop the study. We don't know how another trial would turn out." Raab evidently accepted this advice not to cooperate with the proposed, more rigorous trial. In a compromise, it was agreed that the next TIMI trial would be a test of different angioplasty regimes after tPA treatment—and thus incapable of measuring tPA's efficacy (since all patients received it).[29]

In April 1986, on the basis of the original TIMI study and similar patency data, Genentech filed with FDA for marketing approval, and by the end of the year the firm's leadership were so confident that they expanded the sales force and took out large ads in medical journals proclaiming *"Activase* is Coming." But in May 1987, presented with stronger clinical trial evidence of streptokinase's efficacy from Kabi and Hoechst, which among other things demonstrated survival benefits, the FDA's expert committee on heart drugs voted to support approval of streptokinase for heart attacks but not tPA. An uproar from both the business press and cardiologists ensued—although even the particularly noisy *Wall Street Journal* admitted that the medical world regarded the "delay" as Genentech's own fault. Pressured both by the White House and interested Senators, the FDA approved tPA for sale in November after Genentech submitted evidence from two additional, smallish studies showing that tPA-treated patients had better heart muscle function (measured by ejection fraction) than placebo-treated patients a week after the heart attack. This outcome was of course a compromise between the need to show evidence of clinical benefit and the urgency of Genentech's financial and political backers; the mortality rate of patients after one month or more remained unmeasured.[30]

Activase hit the market with a splash, thanks to the expectations built up around the drug and its maker, registering $60 million in sales in the last two months of 1987 alone. "The cardiologists are salivating," said one hospital pharmacist; also, patients knew all about it from press coverage, and other physicians were enthusiastic too. One reason was Genentech's clever use of clinical trials for marketing purposes, for example testing Activase with emergency room physicians instead of cardiologists. Though meaningless as a trial of tPA's efficacy from a scientific and regulatory perspective, it showed these doctors and hospitals how they could use the drug. Nevertheless, despite the high expectations of physicians and the skill of its marketing team, from the moment of its launch Activase sales suffered from the stiff competition it faced in streptokinase. The price of $2,200, many times higher than stretopkinase, was a particular challenge. In 1988, Activase sales only reached $150 million, about half of what was expected for the first year after such a strong start, and disappointingly levelled off at about $200 million for the next five years.[31]

With the company's hopes pinned on Activase as a blockbuster, Raab's new cadre of experienced pharmaceutical salesmen and marketeers was driven to strain even the drug industry's flexible ethics. For example, in their early Activase marketing campaigns, Genentech not only trained emergency room nurses in using the product, but also paid them to train other nurses and distribute marketing literature in the hospital. In 1991 this "nurse preceptor" program was terminated because it offended doctors and "was starting to smell bad." That same year a Texas man was given tPA for his ongoing heart attack and suffered a stroke; as it later emerged in a lawsuit, he had no option to use streptokinase because Genentech salesmen had (allegedly) offered inducements to hospital pharmacists not to stock alternate drugs. In 1989 when SmithKline Beecham released its competing drug anistreplase or Eminase, a complex of streptokinase with plasminogen extracted from blood, Genentech undertook a national "survey" of doctors asking whether they preferred a pure recombinant product or one made from (potentially infectious) bodily fluids. SmithKline complained that this and other examples of Genentech's aggressive marketing went "beyond the spirit of competition" and fought back with its own limit-stretching sales effort. In April 1994, Genentech scored another dubious marketing first by paying for "continuing medical education seminars" by a physician-lawyer with the message that, even though scientific evidence showed Activase and its competitor thrombolytics to be equally effective, physicians invited malpractice suits if they prescribed a competitor product because Activase was popularly perceived as the standard of care. There were investigations and repeated reprimands from the FDA—reprimands which all drug firms were learning to disregard in the late 1980s and 1990s. But by creatively varying the repertoire, Genentech kept its tPA marketing one step ahead of the law (unlike its growth hormone marketing; chap. 3).[32]

Apart from the price and the fact that it required delivery through a lengthy infusion process, not—as with Eminase and GI's experimental nPA—the convenient single-bolus dose, mounting clinical evidence helped make Activase ever harder for Genentech to sell right through the 1990s. First, in 1990, came news of the test that Grossbard had sought to avoid: a 20,000 patient European mega-trial comparing Activase (supplied by Genentech's European partner Boehringer Ingelheim) against streptokinase for heart attacks. Called GISSI-2, it found no difference in survival rates at 35 days between the two drugs, but it did find that patients treated with Activase suffered a significantly higher risk of stroke. Genentech's drug, in other words, was no more effective than streptokinase in real clinical terms, but more dangerous as well as an order of magnitude more expensive. In 1992 another unwanted (by Genentech), massive 41,000-patient European trial called ISIS-3 compared streptokinase to the tPA that Wellcome manufactured with GI's clones, and to Eminase. All drugs were equal in terms of death rates, but streptokinase again showed a significantly lower rate of bleeding complications than tPA. Anticipating the bad news, Genentech had finally organized its own large comparative trial assessing mortality, saying that the other trials were irrelevant for the US because they did not employ the anticoagulant drug heparin as an adjunct in the exactly same way as most American cardiologists would in practice. Dubbed GUSTO, Genentech's mega-trial compared a new "accelerated" infusion regime of Activase to streptokinase, and to the combination of the drugs, in 40,000 patients. The study found a significant survival advantage for Activase but also a significantly higher stroke rate. In terms of death or nonfatal stroke, the study's main outcome measure, Activase showed a very slight but statistically significant clinical superiority. Thus from the corporate perspective, this expensive—reportedly $50 million—Phase III study was at least a partial success, producing evidence to support claims of superiority that could counter marketing claims by rivals similarly supported by evidence. But despite the best efforts of Genentech's allies in cardiology, experts concluded from this set of large trials that tPA was no more effective than streptokinase in heart attacks, probably a little more dangerous, and that timely treatment was far more important than the choice of drug.[33]

Still, through marketing acumen Genentech managed to retain a strong position among thrombolytic drugs, so it shifted its efforts to promoting the thrombolytic approach generally rather than its own product's superiority. This too was an uphill battle, because the first major clinical trial showing tPA therapy to be inferior to emergency angioplasty was published in 1993, the same year as GUSTO. By the mid-2000s cardiology had left the thrombolytic drugs behind in favor of angioplasty, especially once the original technique of expanding the blocked artery ("balloon angioplasty") was supplemented with the technique of adding a stent or metal tube to keep it open, through the same arm or leg catheter. Thrombolytic drugs as a first line treatment for heart attack have hung on longer in low-tech medical contexts without specialists or the latest imaging technologies. In a 2012 review of the past century's progress by Braunwald, 30 years after Pennica's ovation, tPA did not even merit a mention—although streptokinase and TIMI both did. For all its scientific glamour, Genentech's imagined blockbuster is now barely a footnote in medical history, an innovation that (with streptokinase) played a modest role in the continuous decline that death rates from heart disease have shown since 1980.[34]

By early 1990, when the worrying results from GISSI-2 were beginning to filter out, Genentech was in crisis both financially—having overcommitted to tPA—and in its internal culture. Swanson and Raab had been fighting constantly for five years. Swanson and the board of directors had brought in Raab from Abbott to make Genentech a real drug company with manufacturing, clinical affairs (i.e., trial management), marketing, and sales functions. He had done just that. Although the molecular biologists doing research still enjoyed good working conditions, the firm was far from the informal, egalitarian scientists' republic of ten years before. With the stock price depressed by disappointments like tPA and both men tired of conflict, Roche bought a majority interest, Swanson escaped to the board, and Raab became the chief executive so the firm could focus on building sales revenue (as the directors wanted). Roche took a hands-off stance to running the firm. When Raab was dismissed five years later (according to the *Wall Street Journal,* for bringing multiple lawsuits and investigations by federal agencies and Congress and bad publicity in general upon Genentech, all for "pushing the envelope" in marketing, and also for questionable financial activities), the board installed Art Levinson as chief executive. Levinson was one of the original crew of biologists, so his choice reinforced the firm's science focus and its function as research supplier to Roche. As a Roche executive explained the big firm's rationale in 1997, indirectly referring to these events, the drug industry needed fresh thinking from that "special breed," molecular biologists committed to curiosity-driven research, in order to shore up its dwindling new product pipeline and escape its continuing "me too" (redundant medicine) doldrums. Scientists at Genentech breathed a huge sigh of relief with the management change, seeing Levinson as a "savior" from the "dark ages" of a company culture dominated by extravagantly paid businessmen and, still more distasteful, physician-executives.[35]

Conclusion

Ever the bellwether, Genentech's 1990 takeover by Roche signalled, not the end, but the end of the beginning for the first generation of biotech firms and their products. In 1991, with Genetics Institute stock depressed by Lourie's surprise invalidation of the Hewick Epo patent on appeal (chap. 5), American Home Products, parent company to the drug firm Wyeth-Ayerst, bought a controlling interest. When GI was absorbed by Wyeth it experienced over the next few years a similar culture change to Genentech in its "dark ages"—only permanently. What was left of Genex was taken over in 1991 also, as was Cetus, which had overinvested in interleukin-2 as a potential cancer drug.[36] A handful of the early companies survived a great wave of extinction in the 1990s, more gradually evolving into bigger businesses that still retained a major science focus. Biogen sold its Geneva labs in 1985 under the leadership of Gilbert's replacement, Abbott veteran James Vincent, thus shattering what scientists had experienced as an intense, international research community very like academia. But Vincent retained the active involvement of founding scientific board members including Phil Sharp (who won a life science Nobel in 1993), and Biogen was able to maintain a research orientation through royalties and by selling intellectual property, while marketing just a few products of its own. Similarly Chiron, which brought the original Cetus-developed interferon beta product (Betaseron) to market in 1993, did so by licensing it out to the pharmaceutical firm Berlex. Thus both Biogen and Chiron were able to continue as sites of biotechnology creativity, rather than ordinary drug companies, by only selectively developing products to the market stage, and with those outsourcing manufacturing and many other pharmaceutical operations. The "virtual" pharmaceutical company was pioneered by these firms.[37] An exception that went though no great ruction in the period was Amgen, which became, and remains, a big, fully integrated drug firm. But Amgen, founded by a pharmaceutical executive and always more product-oriented than its biotech competitors, never hosted quite the same type of freewheeling science-driven culture found at Genentech, Biogen, and GI.

By the mid-1990s the science shops of the past decade, still conjured even today by the term "biotech," were things of the past. Certainly there remained, and perhaps still remain, small informal firms where young bench scientists, fresh from academically impressive PhDs and postdocs, reported their own experimental work directly to leading academic scientists—company principals or members of the scientific boards. Maybe there remained, and perhaps still remain, some small firms where young scientists could hope to do important biology, and to publish their work promptly in order to claim due scientific credit. But I do not think there remained any doing all this in the "blue sky" spirit of the early 1980s, freely financed by forgiving investors, where scientists felt ownership in the science while still harboring ambitions to bring drugs to patients and to reap some profit as well. As business scholars have noted, the first wave of biotechs had only enough resources to carry one or two molecules all the way through the lengthy and complex pathways to the medical marketplace. If its patents survived litigation, and its product performed as hoped in clinical trials, and passed FDA muster quickly, and sold very well so as to repay the cost of all those steps and marketing costs too, the biotech survived and became a different kind of organization: a drug firm. If it stumbled along the way, as nearly all did, it was subsumed by a big drug firm or else evaporated with barely a trace. The main options were thus to become a big pharmaceutical company or die trying. The best part for scientists was an ephemeral stage somewhere near the beginning. Biotech's age of innocence, or "naiveté"—as Genentech's founder Boyer put it—was finished.[38]

Science, Business, and Medicine in the First Age of Biotech

As we have seen, the first age of biotech was fuelled by enormous, often unrealistic enthusiasm in both medical and business worlds, and also in politics. The overheated enterprise not only resulted in some excellent science but brought a number of useful new drugs to market, perhaps a few years quicker than they otherwise might have come. In the United States alone, tens of thousands of patients each year still greatly benefit from the drugs discussed in this book. But in the early years especially, many more patients were sold these drugs than benefited. That the first generation of recombinant DNA drugs were oversold does not, of course, take away from their standing as valuable contributions to medicine. It is typical for novel drugs to be initially overused, until the art of medicine discovers a more realistic sense of their value, and this effect was amplified by the excitement surrounding recombinant DNA in the 1980s. Still less does any initial overuse detract from their status as first-rate accomplishments in molecular biology.

The 1980s were something of a golden age, as well as a gold rush, for scientists working to bring the first biotech drugs to market. Of course 1990 was not the end of the era of first-generation recombinant drugs: Epo, which would be the biotechnology sector's best selling product ever, had only been launched on the market the year before. A better landmark would be 2000, the year after Genentech filed for FDA approval of its second generation, single-bolus modified tPA (tenecteplase, branded TNKase), and the year before Amgen filed for its second-generation, long-duration Epo variant Aranesp. But by 1990 the first main cloning races had all been run, the winners were becoming clear as patents were issued and courts decided their scope and disposition, and some second-generation products were in the pipeline. And the finale of the way of scientific life that had so distinctively marked the early biotech era had also begun.

The way of scientific life in the first ten to fifteen years of the biotech firms was, in a word, quasi-academic. In interviews with life scientists who worked at Genentech and Biogen and Genetics Institute in the period, I typically found that their chief motive for taking these jobs was to conduct research under better conditions than otherwise available to them. Some of these young scientists did not perceive their chances of getting an academic post at one of the world's top universities as high, and thought that even a middle-tier university would drag down their research with teaching. Some felt oppressed in postdoctoral positions and wanted more autonomy and credit. Others with a recent PhD did not relish starting the itinerant life of a postdoc, compelled to take an indefinite series of two-year posts wherever they might find them. The pay as well as the working conditions was better than in academia, but nobody recalled that a chance at making big money figured among the reasons for joining the firm. Even allowing for nostalgia, I believe them, because it fits the reality of the day: hitting the jackpot in five or ten years with equity in a tiny firm nobody had heard of (and with present value essentially nil) was never a likely enough payoff to justify, by itself, staking the best years of your life. Still, equity ownership was an attraction, because it provided young biologists both material and symbolic investment in their work. So at least in these three firms, scientists were basically there to do life science, and they still sought scientific credit for their work by publishing promptly. Some did chafe at what they experienced essentially as a well-paid postdoctoral role, when at that stage of their career they might have had the control over their research enjoyed by a junior professor. But so long as they were reporting directly to accomplished scientists whom they genuinely respected, the trade-off between autonomy and quality of science was acceptable.[1]

For a while, all this was really possible. The eminent academic biologists who started firms and staffed the early scientific boards talked to their scientists directly, not through salaried managers. Because of where the frontiers of molecular biology were situated in the 1980s, isolating a human gene from cDNA or a genomic library, and cloning and sequencing it, and expressing it in a novel genetically engineered construct represented important scientific challenges. Such accomplishments were publishable in the best journals. And the academic biologists who were company principals generally wanted their firms, and therefore themselves, to share in the scientific laurels that could come from prominent publications by their scientists. Commingled with the prestige, publications in *Science* and *Nature* were also a way of signalling claims to intellectual property long before the patent situation could be assessed (especially in the pre-2000 US patent regime, where patents were published not on application but only upon issue), and therefore useful to impress investors as well as other scientists. And so long as the initial investors' money lasted—and then the early IPO takings—price was no object in winning the cloning races that would also win the publications and patents and credit.

In the labs at the top biotech firms, the young molecular biologists worked in ways closely parallel to their prior experience in the world's best university labs, while pursuing similar scientific projects. In effect these firms transplanted an ideal academic culture—idealized in the sense of carrying on the teaching-free basic research ideology under new commercial flag, and also in the sense of naively detached from what the drug business entailed. For example, they reproduced existing tribal divisions among academic molecular biologists, as evidenced by company departments defined by names like Molecular Genetics and Protein Chemistry, within which there were "lab chiefs" and much the same the informal job hierarchy as in an academic lab. Their fierce competitiveness to be the first to clone gene A or solve structure B, and to claim credit whether through patent or publication—and typically both—was essentially the same as already existed in the world of academic molecular biology. (Indeed William Rutter later reflected that academic life science, entirely driven by the quest for reputation, was actually more competitive than business.)[2]

Biologists managing and working in firms did revise established behavioral codes, but not so dramatically as often supposed. We have seen how patent inventorship immediately became assimilated to existing systems of credit by publication authorship (chap. 3). For another example, in announcing their accomplishments biotech scientists began claiming public credit with press conferences long before publication, attracting criticism for bypassing peer review. But even so, in the firms I studied they generally maintained scientific credibility by presenting immediately beforehand to peers at academic seminars—symbolically placing science first. While competition between labs was often ferocious already, as I have just argued, they may have raised it one more notch (perhaps from "ten" to "eleven," so to speak). Thus they learned to excuse, or at least rationalize, previously unsporting behavior like setting up cameras to photograph competitors' academic presentations. And while camping out in labs to continue experiments overnight was already a university commonplace, they may have accelerated an already cracking pace of work because, even more so than in academia, the cloners were all pursuing essentially identical projects, all racing to harvest the same "low hanging fruit." Offered spontaneously in interviews by many biologists who then worked in biotech, this term may or may not have been used the same way circa 1980. But the concept was certainly there: as Cetus President Farley complained as early as 1979, "everyone has the same damn list" of cloning targets. One biologist spontaneously defined the low-hanging fruit thus: "[physiologically active] proteins that there was sufficient amount of biology . . . already known about them that people thought . . . if you could make enough of them, you could turn them into drugs." The prime cloning targets on the shared list of low hanging fruit were on that "same damn list" precisely because they were *already* subjects of intense and longstanding interest in the world of biomedical science. That is why so much was already known. No wonder that the first commercial races to harvest them attracted top-notch life scientists, and that accomplishments were fully fungible for academic credit— as evidenced both by publications in *Science* and *Nature*, and the later academic careers of alumni from firms like Genentech, Biogen, and Genetics Institute.[3]

Since this book has focussed on the science and scientists within the firms, it cannot offer the same degree of resolution about what was simultaneously happening in the nearby world of academic molecular biology. But insights do emerge. The new business ambitions of entrepreneurial professors may well have heightened competitiveness and especially secretiveness on the academic side of the porous border, as commentators at the time suggested. When they considered research topics, among the genes and gene products presenting intellectual opportunity at the time, they may well have focussed their academic work upon projects which—even though entirely justifiable and important scientifically—happened also to offer the greatest commercial promise. The value system within academic molecular biology expanded so as to include essentially commercial outcomes, like patents and instrumental steps toward large-scale production, as accomplishments proper to biology professors and academic labs, and to forgive "cloning by press conference" if the subsequent publication delivered the goods. Whatever the main reasons for the NIH guideline violations in his UCSF lab and their tardy disclosure, Rutter revealed how the commercial race for intellectual property and the academic competition for credit melded when in 1977 he described a motivation (albeit one outweighed by fear of stifling regulation) as securing "the supremacy of our cloning experiments" (chap. 2). Even academic life scientists with no direct financial interest in a biotech company began to see commercial reward as just desserts for a record of distinguished research, as when the NIH erythropoietin distribution panel declined to supply the protein to Biogen-affiliated academic scientists because of Goldwasser's contract with Amgen.

To be sure, some university biologists wedded to the ideology of pure research that prevailed in the early Cold War did sound alarms at the commercial turn, and were echoed by early critics in the social sciences. Indeed, in the late 1970s and early 1980s, some entrepreneurial biologists were punished by academic colleagues for perceived commercial transgressions. But for many if not most biologists, so long as there seemed no inconsistency between doing (and publishing) top-level biology, bringing medically useful new drugs closer to the bedside, and trying to make money, it was possible to pursue all three projects at once without losing scientific integrity and credibility.[4] It was much the same among the endocrinologists and biochemists half a century earlier, when the resonance between intellectual, medical, and business opportunities around hormones led them to remold the pure science ethos of their day. Many of these hormone researchers won Nobel prizes for the work they did in collaboration with industry in the 1930s and 1940s. So did quite a few of the molecular biologists involved with biotech firms in the 1980s and 1990s, although not always for their commercial accomplishments.

While it is certainly a grave matter for the United States that the nation has committed ever-diminishing public funding to science since the period on which this book focuses (from 1.27% of gross domestic product in 1976 to 0.87% in 2011, one-third less), it is surely unjust to blame the academic biologists who first felt the chill winds and found ways to carry on doing good science with private support. If the nation wishes scientists to pursue curiosity-driven research with less immediate commercial potential, then its political leadership must find a way to restore the funding that gives scientists that option. And while this book offers reasons to suppose that the record of scientific productivity established in the first age of biotech cannot be repeated, because all the low hanging fruit is now harvested, it does suggest policy lessons. For one, the dependence of that early productivity on publicly funded basic biology in the 1970s implies that, moving into the future, better public support of curiosity-driven biology is needed to further the productivity of biotechnology enterprise, and perhaps less emphasis on rapid payoff.[5]

The heady amalgam of academic science and business interests that characterized biotech's first age was inherently unstable, held together by the coincidence of molecular biology's intellectual and methodological frontier with business opportunities. Those business conditions were the product of an even more ephemeral situation outside science's imagined boundaries. Entrepreneurial academic biologists in fact played a large role in generating that fertile situation, for example by forestalling legal restrictions on recombinant DNA by voluntarily regulating themselves through the NIH guidelines, by stoking and harnessing public hopes in interferon as a cancer cure and genetic engineering as an economic elixir, and of course by as acting as emissaries and salesmen to the investment world. A pervasive enthusiasm for linking universities to the private sector, associated with the dawn of neoliberalism, made their nascent ventures seem especially promising and exciting.

The happy conjunction of their intellectual interests as biologists and their financial interests as entrepreneurs was destabilized by events beyond their control in the early 1990s—even while the scientific frontier had yet not passed beyond what was useful to develop medical products (for example the reconstitution of eukaryotic transcriptional regulatory elements in expression plasmids, or the predictive protein engineering involved in making second-generation recombinant drugs). By then the enormous medical hopes in interferon, so crucial to the massive stock offerings that floated the first biotechs in the early 1980s, had vanished. A drought of investment capital struck around 1994, related to macroeconomic conditions, darkening the biotech skies. As I have shown, it was also in the early 1990s that the winners of the cloning races for the low-hanging fruit, and therefore the few companies most likely to offer a first-generation drug, were becoming clear through the issue of US patents, announcements of clinical trials, and the early decisions of courtroom infringement battles.[6] And as I have also argued (chaps. 5, 6), also crucial was the simultaneously developing shape of patent law—anticipated court decisions built into business thinking, as well as actual precedent and written statute—that by the early 1990s was altering the rationale and incentives for investment in small biotech firms.

Over time, the established drug industry's lack of competence in molecular genetics passed, and the big companies found themselves less reliant on the specialized knowledge possessed only by academic experts and their biotech start-ups. In the early days, when big drug firms didn't know much more about recombinant DNA than FDA Commissioner Kennedy jokingly professed, it had been possible to found a company on the premise that it could supply basic knowledge that drug firms would pay for, and/or win broad patents on methods that might either be sold or profitably licensed. Whole firms seem to have been built and sold on this premise that "blue-sky" life science could make money. DNAX, for instance, was seen by the business world as a "biology think tank." The very broad and early Cohen-Boyer patents encouraged such belief, and through reasonable license fees did indeed generate $200 million over their lives for Stanford and the University of California. Columbia University's early patents on basic methods for expressing cloned genes in cultured mammal cells would garner that university nearly $800 million. Riggs' and Itakura's patents on expression vectors employing synthetic coding sequences earned City of Hope hundreds of millions from its licensing to Genentech, and also enabled Genentech to hobble competition through its monopoly on the method.[7]

But amid the rush of overlapping applications and interference proceedings, the US patent office in the later 1980s understandably issued narrower patents. As we have further seen, especially in chapter 5, the US appeals court (CAFC) hearing most of the important biotechnology infringement lawsuits made a series of decisions that supported exclusive protection on first generation products for those who cloned natural genes, but restricted their scope so as to exclude second-generation variants. They simultaneously set a very low bar on the inventiveness (i.e., nonobviousness) required for patent protection. The combination of intermediate scope of patent protection offering monopolies on first-generation natural protein drugs regardless of their mode of production, and no great reward for inventiveness, made it less plausible to envisage payoffs from financing relatively fundamental biology research. This court-driven shift in the perceived rewards for biotechnology investment in the early 1990s coincided with the declaration of the winners in the first cloning races (often in the same court decisions). So the low-hanging fruit were taken too. Thus, even apart from macroeconomic conditions, the interaction between biology and law meant that financing for biotechs inevitably became harder in the early 1990s.

These same early 1990s developments in patent case law changed the complexion of the science that the biotechs needed to do to stay in business. As I have argued, the CAFC's biotechnology decisions just alluded to, in offering strong protection of first generation products irrespective of inventiveness (i.e., nonobviousness), explicitly encouraged the development of second-generation "me-too" variants of relatively low inventiveness and novelty—at least in terms of therapeutic effect—along the same lines as the established drug industry. To continue the more academically inspired basic research with which many biotech enterprises began, the sector would have needed the courts to support broad patents, with an emphasis on inventiveness of method or mode of discovery, in a way they did not. Biotech firms had little alternative but to adapt to this me-too incentive scheme, which was already the path of least resistance because of a basic feature of the drug business: the inherently greater ease of clinical testing and speed of approval for fairly well-understood biomedical entities.

But there was nothing inevitable or obviously optimal about this de facto science policy that the US appeals courts developed and imposed through case law. German chemical and pharmaceutical firms thrived in the late nineteenth century and came to dominate the world through patent law that only protected manufacturing methods, but not medicines or other chemicals as products. Switzerland, with the second most powerful pharmaceutical industry in the early twentieth century, had no patent law at all. Indeed, only Germany's defeat in the World War I (wherein German firms' obstructive US product patents were seized and redistributed) allowed American research-based firms a real start. As I have pointed out, the British High Court gestured in the direction of methods-driven, product-by-process protection in its tPA decision, reasoning that the uninventiveness of Genentech's natural protein and the obviousness of the way it was cloned did not merit broad protection. Thus, the greater the inventiveness of method of isolation or production, the broader the protection a product deserved. But the CAFC chose an opposite policy, essentially disregarding arguments about the obviousness of methods and focusing reward on products, extending protection broad enough to secure monopolies on natural first-generation proteins. Patent courts in Europe and Japan, and soon the UK, all generally shifted in the US direction—influenced no doubt by the image of a revolutionary, booming American biotech sector, not to mention the growing power of certain multinational drug firms to bend global trade policy to their demands (as distinct from the drug industry's long-term needs).[8] Essentially, alternate intellectual property arrangements to maximize economic and medical benefit from modern biotechnology have not been explored.

As a matter of record, the economic performance of the biotechnology sector in the era of the first generation drugs did not live up to the contemporary notion of a boom—let alone to the extraordinary early expectations circa 1980, when a revolutionary new industry was expected to restore America with a magic economic elixir. Talk of molecular biology visiting Schumpeterian "creative destruction" upon the pharmaceutical sector is now rare. According to one influential economic analysis, the collective business performance of biotechnology firms in the late 1980s through the early 2000s just equaled that of traditional drug firms. Furthermore, perhaps pointing toward a grimmer future, if the analysis excludes Genentech and Amgen, biotech performed worse than pharmaceuticals more generally—and the analysis that includes them happens to cut off right before sales of the biggest product in the entire sector, erythropoietin, plummeted by billions in recognition of the drug's widespread, deadly overuse (chap. 5). Another influential analysis of biotechnology's productivity, measured in terms of drug innovation rather than sales, found that overall new drug submissions for regulatory approval have not increased since biotech firms started producing them, and that a biotech-assisted wave of new drug patents has only kept up with the skyrocketing research expenses of the flagging pharmaceutical sector. Thus there is strong evidence indicating that talk of a biotechnology revolution, whether viewed through an economic or medical lens, is a great exaggeration if not a myth. Great science aside, biotechnology pharmaceuticals development has translated into pharmaceutical business as usual, quantitatively speaking.[9]

All this is exactly what one expects if, as I have argued in this book, biotech was quickly forced by American legal as well as financial and regulatory institutions to become simply an ordinary "me-too" part of the drug industry. But even apart from that contention, if I am right in my educated guess that new products will from now on come much slower and harder, because of the passing of the scientific era of low-hanging fruit that became the first-generation recombinant drugs, then we can never again expect the measure of business success so far recorded. No doubt, so long as there are entrepreneurial biologists, there will always be a few winners with genuine biomedical breakthroughs for the lucky investor to pick. Yet any efforts to reproduce the remarkable, creative atmosphere of the first two decades when molecular biology met the business world—dependent as it was on the conjunction of ephemeral conditions in science, culture, law, and business—would be as futile as capturing a breaking wave in a bottle.

Abbreviations

BLR: Biotechnology Law Report
JAMA: Journal of the American Medical
  Association
JBC: Journal of Biological Chemistry
JHB: Journal of the History of Biology
JMB: Journal of Molecular Biology
NBT: Nature Biotechnology

NEJM: New England Journal of Medicine
NYT: New York Times
PNAS: Proceeding of the National Academy
  of Sciences, USA
SSS: Social Studies of Science
WSJ: Wall Street Journal

Interviews

Interviews by the Author

Rasmussen-Davies: Julian Davies,
  October 2011
Rasmussen-Fritsch: Edward Fritsch,
  May 2011
Rasmussen-Gilbert: Walter Gilbert,
  August 2011
Rasmussen-Hall: Alan Hall, June 2011
Rasmussen-Kay: Robert Kay, April 2012
Rasmussen-Larsen: Glenn Larsen,
  April 2012
Rasmussen-Lomedico: Peter Lomedico,
  August 2011
Rasmussen-Maniatis: Tom Maniatis,
  November 2010

Rasmussen-Nagata: Shigekazu Nagata,
  December 2011
Rasmussen-Rosa: Joseph Rosa,
  April 2012
Rasmussen-Seeburg: Peter Seeburg,
  March 2011
Rasmussen-Shine: John Shine,
  February 2011
Rasmussen-Vapnek: Daniel Vapnek,
  January 2012
Rasmussen-Villa-Komaroff: Lydia Villa-
  Komaroff, May 2011

Interviews by Sally Hughes (available at Regional Oral History collection of the University of California, Bancroft Library, Berkeley; available online at time of writing, http://bancroft.berkeley.edu/ROHO/projects/biosci/oh_list.html)

Hughes-Boyer: Herbert Boyer, 1994
Hughes-Cape: Ronald Cape, 2003
Hughes-Goeddel: David Goeddel,
   2001 and 2002
Hughes-Heyneker: Herbert Heyneker,
   2002
Hughes-Itakura: Keiichi Itakura, 2005
Hughes-Johnson: Irving Johnson, 2004
Hughes-Kiley: Thomas Kiley, 2000 and
   2001
Hughes-Kleid: Dennis Kleid, 2001 and
   2002

Hughes-Pennica: Diane Pennica, 2003
Hughes-Rathmann: George Rathmann,
   2003
Hughes-Riggs: Arthur Riggs, 2005
Hughes-Rutter: William Rutter, 1992
Hughes-Swanson: Robert Swanson,
   1996 and 1997
Hughes-Ullrich: Axel Ullrich, 1994 and
   2003
Hughes-Yansura: Daniel Yansura, 2001
   and 2002
Hughes-Young: Bill Young, 2004

Court Case Records, Federal District Courts of the United States

Amgen v. Chugai files: Records of Amgen v. Chugai and Genetics Institute, CA, No. 87-2617-Y (Mass. Dist.) and associated countersuits and appeals (accession AAC1-10743596) NARA Federal Records Storage Facility, Waltham, Massachusetts.

Amgen v. Elanex files: Records of Amgen v. Elanex, CA, C93-1483D (W. Dist. Wash.), and associated countersuits and appeals (Accession 021-00-0026), NARA Federal Records Storage Facility, Seattle, Washington.

BTG v. Genentech files: Records of Bio-Technology General v. Genentech, 95 CV 0110 (So. Dist. NY) and associated countersuits and appeals (Accession AAC1-17626179), NARA Federal Records Storage Facility, Lee's Summit Missouri.

Genentech v. Boehringer files: Records of Genentech v. Boehringer-Mannheim, CA, 96-11090-PBS (Mass. Dist.), and associated countersuits and appeals (Accession 021-03-0023), NARA Federal Records Storage Facility, Waltham Massachusetts.

Genentech v. Wellcome files: Records of Genentech v. Wellcome and Genetics Institute, CA, 88-330-JJF (Del. Dist.), and associated countersuits and appeals (accession AAC1-349492), NARA Federal Records Storage Facility, South Philadelphia.

Schering v. Amgen files: Records of Schering and Biogen v. Amgen, CA, 96-587 MMS (Del. Dist.), and associated countersuits and appeals (Accession AAC1-183216), NARA Federal Records Storage Facility, South Philadelphia.

Introduction · Biology's Day at the Races

1. Rai AK and Eisenberg RS, Bayh-Dole reform and the progress of biomedicine, *Law Contemp Probl* 2003;66:289–314; Mowery D and Sampat B, The Bayh-Dole Act of 1980 and university industry technology transfer: a model for other OECD governments? *J Technol Transfer* 2005;30:115–27; Pisano G, *Science business: the promise, the reality, and the future of biotech* (Cambridge, MA: Harvard Business School Press, 2006); Hopkins M, Martin P, Nightingale P, et al., The myth of the biotech revolution: an assessment of technological, clinical and organisational change, *Res Policy* 2007;36:566–89.

2. I refer among other things to enormous duplication of effort and the costly litigation noted in several chapters, to monopoly profits even greater than standard in pharmaceuticals due to Orphan Drug exclusivity on blockbuster products, and above all to the many thousands of deaths caused by the medically inappropriate overuse of erythropoeitin. The sales of this one drug were so great that their assessment plays a key role in the biotechnology sector's overall economic record. See discussions of Epogen, Aranesp, and Procrit in chap. 5. Other best-selling first-generation biotechnology drugs also owe substantial sales to marketing-driven, medically inappropriate, and to some extent harmful use, but probably not much more so than successful new pharmaceuticals generally.

3. Durkheim E, *On the division of labor in society*, 2nd ed. (1893), Simpson G, trans. (New York: Macmillan, 1933), preface.

4. For excellent examples of this genre, see Kenney M, *Biotechnology: the university-industrial complex* (New Haven: Yale University Press, 1986); Krimsky S, *Biotechnics and society: the rise of industrial genetics* (New York: Praeger, 1991); Kleinman D, *Impure cultures: university biology and the world of commerce* (Madison: University of Wisconsin Press, 2003). Among the many early books in the business literature celebrating great benefits from biotech, one of the most valuable is Orsenigo L, *The emergence of biotechnology* (Cambridge, MA: Harvard University Press, 1989).

5. Mirowski P, *Science mart: privatising American science* (Cambridge, MA: Harvard University Press, 2011). Also see Kleinman D, *Impure cultures: university biology and the world of commerce* (Madison: University of Wisconsin Press, 2003), and Krimsky S, *Science in the private interest* (Lanham, MD: Rowman-Littlefield, 2003). For a compatible, broader critique of neoliberal intellectual property and science policy, see Jaffey AB and

Lerner J, *Innovation and its discontents: how our broken patent system is endangering innovation and progress, and what to do about it* (Princeton: Princeton University Press, 2011).

6. Mirowski, *Science mart.* I resist a tragic emplotment because I do not see the molecular biology of the Cold War era as divorced from the drug industry. There was simply a different division of labor, and one not necessarily better for the public health or society generally. On historians and their narrative choices, see White H, Historical emplotment and the problem of truth in historical representation. In: *Figural realism: studies in the mimesis effect* (Baltimore: Johns Hopkins University Press, 1999):27–43.

7. Ducor P, *Patenting the recombinant products of biotechnology and other molecules* (London: Kluwer, 1998).

8. There is one way in which the typicality of the five drugs selected here may be questioned. I have not studied two of the first recombinant drugs whose genuinely novel contribution to therapeutics has arguably been especially valuable, granulocyte colony stimulating factor (introduced by Amgen and used for boosting immune function in cancer chemotherapy patients) and interferon beta (introduced by Cetus/Chiron and Biogen, and used for retarding multiple sclerosis). In the latter case I decided to study the earliest of the interferons developed; in the former, no useful lawsuit materials could be found. The reader should thus observe caution in concluding from this study that no early recombinant drugs made a unique and novel contribution to medicine greater than the five drugs discussed here. Further empirical analysis would be required for this argument.

9. While I have relied extensively on Hughes's interviews, her corporate history of Genentech was published very late in the writing of this book and was not read before I completed a full draft. See Hughes S, *Genentech: the beginnings of biotech* (Chicago: University of Chicago Press, 2011).

10. On "gentlemen" and patent litigation in the middle twentieth century drug industry, see Rasmussen N, *On speed: the many lives of amphetamine* (New York: New York University Press, 2008), chap 4.

Chapter 1 · Biology, Industry, and the Cold War

1. Boyer P. *By the bomb's early light: American thought and culture at the dawn of the atomic age* (New York: Pantheon, 1985):156; radio excerpt from "WEAF, the national hour (as broadcast)," script attached to Niles Trammel to Warren, January 11, 1946, Stafford Warren Personal Papers, Special Collections of the University Research Library, University of California at Los Angeles, collection RG 987, box 285, folder "Speeches." Weart S, *Nuclear fear: a history of images* (Cambridge, MA: Harvard University Press, 1988); Keller EF, Physics and the emergence of molecular biology: a history of cognitive and political synergy, *JHB* 1990;23:389–409; Rasmussen N, The mid-century biophysics bubble: Hiroshima and the biological revolution in America, revisited, *Hist Sci* 1997;35:245–93.

2. An April 2010 full text search for "molecular biology" in Proquest Historical Newspapers, comprising the entire content of the *Boston Globe, Chicago Tribune, Christian Science Monitor, Los Angeles Times, New York Times,* and *Wall Street Journal* returned no hits for the 1930s and only four stories for the 1940s. In contrast "biophysics" returned 70 stories for all of the 1930s, 31 stories for 1940–45, and 169 items for the four years 1946–49 alone. Full-text searches of *Science* and *Nature* magazine each find only one use of the term "molecular biology" before 1950, and these by Weaver-sponsored researchers Doro-

thy Wrinch and William Astbury; Notes and news, *Science* 1941;93:591; Astbury W, Progress of X-ray analysis of organic and fibre structures, *Nature* 1946;157:121–24. The term only becomes common in the later 1950s. The impression that molecular biology existed before 1950, in the mind of anyone but Weaver and his closest associates, may be due to the lavish attention historians have focused on Weaver—as well as the general anachronistic tendency to read this disciplinary concept's later success into the deeper past.

3. Pauly P, *Controlling life: Jacques Loeb and the engineering ideal in biology* (New York: Oxford University Press, 1987); idem, General physiology and the discipline of physiology, 1890–1935. In: Geison G, ed., *Physiology in the American context 1850–1940* (Baltimore: American Physiological Society, 1987):195–207; Schwerin A, Prekäre stoffe: radiumökonomie, risikoepisteme und die etablierung der radioindikatortechnik in der zeit des nationalsozialismus, *Zeitschrift für Geschichte der Naturwissenschaften, Technik und Medizin* 2009;17:5–33; Summers WC, Biophysics in Berlin: the Delbrück club; and Sloan P, Exhuming the three-man paper: target-theoretical research in the 1930s and 1940s. Both in Sloan P and Fogel B, eds., *Creating a physical biology: the three-man paper and early molecular biology* (Chicago: University of Chicago Press, 2011):39–60, 61–98.

4. Olby R, *The path to the double helix* (Seattle: University of Washington Press, 1974); Abir-Am P, The discourse of physical power and biological knowledge in the 1930's: a reappraisal of the Rockefeller Foundation's "policy" in molecular biology, *SSS* 1982;12:341–82; Kohler R, *Partners in science: foundations and natural scientists, 1900–1945* (Chicago: University of Chicago Press, 1991); Kay LE, *The molecular vision of life: Caltech, the Rockefeller Foundation, and the rise of the new biology* (New York: Oxford, 1993); Weaver W, Molecular biology: origin of the term, *Science* 1970;170:581–82; Rasmussen, Midcentury biophysics bubble.

5. Weaver W, *Rockefeller Foundation annual report*, 1934, Rockefeller Archive Center, Tarrytown New York: 198–99. See Kay, *Molecular vision*, chap. 1.

6. Roll-Hansen N, The progress of eugenics, *Hist Sci* 1988;26:295–331; Kay LE, Life as technology: representing, intervening and molecularizing, *Rivista di Storia della Scienza* 1993;1:85–103; Allen G, The social and economic origins of genetic determinism: a case history of the American Eugenics Movement, 1900–1940 and its lessons for today, *Genetica* 1997;99:77–88; Paul, D, *The politics of heredity: essays on eugenics, biomedicine, and the nature-nurture debate* (Albany: SUNY Press, 1998); Kevles D, Eugenics, the genome, and human rights, *Medicine Studies* 2009;1:85–93. Historians who argue that American geneticists distanced themselves from eugenics, based on their scientific writings, often overlook the extent that they still took advantage of ongoing eugenics enthusiasms in the 1930s (including among Rockefeller Foundation trustees). Gleason S, Sensational study of heredity may produce new race of men, *Popular Science*, November 1934, 15–17.

7. Oudshoorn N, *Beyond the natural body: an archaeology of the sex hormones* (London: Routledge, 1994):88; Rasmussen N, The moral economy of the drug company–medical scientist collaboration in interwar America, *SSS* 2004;34:161–86; Kohler RE, *Lords of the fly: Drosophila genetics and the experimental life* (Chicago: University of Chicago Press, 1994).

8. Haynes W, *American chemical industry: decade of new products, 1930–1939* (New York: Van Nostrand, 1954), chap. 18; Corner G, The early history of the oestrogenic hormones, *J Endocrinol* 1965;33:3–17; Parkes A, The rise of reproductive endocrinology, *J Endocrinol* 1966;34:20–33; Weiner C, Patenting and academic research: historical case

studies, *Sci Technol Hum Val* 1987;12:50–62; Apple R, Patenting university research: Harry Steenbock and the Wisconsin Alumni Research Foundation, *Isis* 1989;80:375–94; Surgenor D, *Edwin Cohn and the development of protein chemistry* (Boston, MA: Center for Blood Research, 2001), chap. 3; Gaudilliere J-P, Better prepared than synthesized: Adolf Butenandt, Schering AG and the transformation of sex steroids into drugs 1930–1946, *Stud Hist Phil Biol Biomed Sci* 2005;36:612–44; Rasmussen, Moral economy, and idem, Steroids in arms: Science, government, industry, and the hormones of the adrenal cortex in the United States, 1930–1950, *Med Hist.* 2002;46:299–324; THEELIN, Female Sex Hormone in Crystalline Form, Now Available (advertisement), *Am J Med Sci* 1931;181:24.

9. On the mention rate and connotations of "biophysics" in major news media, see n. 2 above. The earliest explicit description of microbial genetics as "biophysics" (or "molecular biology") comes from late 1948 and refers to Leo Szilard; Gibbons R, Virus offspring are mated in heredity study, *Chicago Tribune,* October 25, 1948:14.

10. Feffer S, Atoms, cancer, and politics: supporting atomic science at the University of Chicago, 1944–1950, *Hist Stud Phys Biol Sci* 1992;22:233–36; Rasmussen, Biophysics bubble; Creager A, Radioisotopes as political instruments, 1946–1953, *Dynamis* 2009;29:219–39.

11. Rasmussen, Biophysics bubble; Kay, *Molecular vision;* Creager A, Wendell Stanley and the dream of a free-standing biochemistry department at the University of California, Berkeley, *JHB* 1996; 29:331–60; Muller, H, Artificial transmutation of the gene, *Science* 1927;66:4–87; Muller, H, The production of mutations, *Nobel lectures in physiology or medicine, 1942–1962* (New York: Nobel Foundation, 1967):154–71; Kay LE, W.M. Stanley's crystallization of the Tobacco Mosaic Virus, 1930–1940, *Isis* 1986; 77:450–72.

12. Friedberg EC, A brief history of the DNA repair field, *Cell Res* 2008;18:3–7; Kay LE, Conceptual models and analytical tools: the biology of physicist Max Delbrück, *JHB* 1985; 18:207–46; Hayes W, Max Ludwig Henning Delbrück, 4 September 1906–10, March 1981, *Biographical Memoirs Fellows Roy Soc* 1982;28:58–90.

13. Ravin A, The gene as catalyst, the gene as organism, Stud Hist Biol 1977;1:1–45; Olby, *Path;* Morange, M, *A history of molecular biology,* trans. Cobb M (Cambridge MA: Harvard University Press, 1998).

14. Morange, *History,* chap. 12; Rasmussen N, Mitochondrial structure and the practice of cell biology in the 1950s, *JHB* 1995;28:381–429; Rheinberger H-J, *Toward a history of epistemic things: synthesizing proteins in the test tube* (Stanford: Stanford University Press, 1997).

15. Kornberg A, Enzymatic synthesis of deoxyribonucleic acid, *Harvey Lectures* 1959;53:83–112; Lehnman, IR, DNA ligase: structure, mechanism, and function, *Science* 1974;186:790–97.

16. Chargaff E, How genetics got a chemical education, *Ann NY Acad Sci* 1979;325:345–60; Olby R, Biochemical origins of molecular biology: a discussion, *Trends Biochem Sci* 1986;11:303–5; Crick F, Looking backwards: a birthday card for the double helix, *Gene* 1993;135:15–8; Fruton J, *Proteins, enzymes, genes: the interplay of chemistry and biology* (New Haven: Yale University Press, 1999), chap. 7. On the "pidgin" concept see Galison P, History, philosophy, and the central metaphor, *Science in Context* 1988;2:197–212.

17. Pauling L, Itano HA, Singer SJ, Wells IC, Sickle cell anemia: a molecular disease, *Science* 1949;110:543–48; Strasser B, Perspectives: sickle cell anemia; a molecular disease,

Science 1999;286:1488–90; Schechter AN, Christian B, Anfinsen 1916–1995, *Nat Struct Biol* 1995;2:621–23; Fruton, *Proteins, enzymes, genes*, chap. 5.

18. Yanofsky C, Gene structure and protein structure, *Harvey Lect* 1967;61:145–68; Judson, H, *Eighth day of creation: makers of the revolution in biology* (New York: Simon & Schuster, 1979), chap. 6; de Chadarevian S, Sequence, conformation, information: biochemists and molecular biologists in the 1950s, *JHB* 1996;29:361–86; Morange, *History*, chap. 15; Fruton, *Proteins, enzymes, genes*, chap. 7; Kay, LE, *Who wrote the book of life? a history of the genetic code* (Stanford: Stanford University Press, 2000), chap. 4; Baltimore D, Discovery of the reverse transcriptase, *FASEB J* 1995;9:1660–63.

19. Judson, *Eighth day*, chap. 8; Rheinberger, *Toward a history*, chaps. 10–13; Morange, *History*, chaps. 12–13. Historians have focussed on the NIH group of Marshall Nirenberg and largely neglected the simultaneously published work of Severo Ochoa's New York group; see Lengyel P, Speyer JF, and Ochoa S, Synthetic polynucleotides and the amino acid code, *PNAS* 1961;47:1936–42.

20. Morange, *History*, chaps. 5, 14; Kay, *Who wrote the book of life*, chap 5. On the early linkage of virus studies to cancer also see Galperin C, Virus, provirus, et cancer, *Rev Hist Sci* 1994; 47:7–56; Creager A, *The life of a virus: Tobacco Mosaic Virus as an experimental model, 1930–1965* (Chicago: University of Chicago Press, 2001).

21. Ibid.; Judson, *Eighth day*, chap. 7; Jacob F and Monod J, Genetic regulatory mechanisms in the synthesis of proteins, *JMB* 1961;3:318–56.

22. Morange, *History*, chap. 15; idem, The transformation of molecular biology on contact with higher organisms, 1960–1980: from a molecular description to a molecular explanation, *Hist Phil Life Sci* 1997;19:369–93.

23. Caspersson T, Award presentation speech, 1958 Nobel Prize in physiology or medicine (http://www.nobelprize.org/nobel_prizes/medicine/laureates/1958/press.html) (quote); Kay LE, Selling pure science in wartime: the biochemical genetics of G.W. Beadle, *JHB* 1989;22:73–101; Novick A and Szilard L, Experiments with the chemostat on spontaneous mutations of bacteria, *PNAS* 1950;36:708–19; Franck J and Gaffron H, Research in progress-the Institute of Radiobiology and Biophysics, in Research in progress reports of May and October 1953, Presidents Papers 1950–55 and 1952–60, University of Chicago Archives, box 141, folder 141.2; Zirkle R, Biophysics at the University of Chicago, unpublished presentation, in "Conference on status of biophysics, Ann Arbor Michigan," September 16, 1955, Francis Schmitt papers, Massachusetts Institute of Technology archives, carton 21, folder 12; Rasmussen, Biophysics bubble; Chakrabarty, AN and Kellogg ST, Bacteria Capable of Dissimilation of Environmentally Persistent Chemical Compounds, US Patent 4,535,061, issued August 13, 1985, on application of December 28, 1981.

24. Rasmussen, Midcentury biophysics bubble; idem, *Picture control: the electron microscope and the transformation of biology in America, 1940–1960* (Stanford: Stanford University Press, 1997); de Chadarevian S. *Designs for life: molecular biology after World War II* (Cambridge: Cambridge University Press, 2002); Nye M-J, Paper tools and molecular architecture in the chemistry of Linus Pauling, *Boston Stud Phil Sci* 2001;222:117–32; Gavroglou K and Simões A, *Neither physics nor chemistry: a history of quantum chemistry* (Cambridge MA: MIT Press, 2012); Strasser B, Institutionalizing molecular biology in post-war Europe: a comparative study, *Stud Hist Phil Biol Biomed Sci* 2002;33:533–46; Appel T, A shaping biology: the National Science Foundation and American biolog-

ical research, 1945–1975 (Baltimore: Johns Hopkins University Press, 2000), 207–34; Krige J, The Birth of EMBO and the difficult road to EMBL, *Stud Hist Phil Biol Biomed Sci* 2002;33:547–64.

25. Merton RK, *The sociology of science: theoretical and empirical investigations* (Chicago: University of Chicago Press, 1973):267–80 et passim; Kuhn T, *The structure of scientific revolutions* (Chicago: University of Chicago Press, 1962); Hollinger DA, Science as a weapon in *Kulturkampfe* in the United States during and after World War II, *Isis* 1995;86:440–54; Fuller S, *Thomas Kuhn: a philosophical history for our times* (Chicago: University of Chicago Press, 2000); Shapin S, *The scientific life: a moral history of a late modern vocation* (Chicago: University of Chicago Press, 2008), chap. 3.

26. Figures from Baxter JP 3rd, *Scientists against time* (New York: Little-Brown, 1946):300. Geiger R, What happened after Sputnik? shaping university research in the United States, *Minerva* 1997;35:349–67; Kleinman D, *Politics on the endless frontier: postwar research policy in the United States* (Durham: Duke University Press, 1995); Appel TA, *Shaping biology: the National Science Foundation and American biological research, 1945–1975* (Baltimore: Johns Hopkins University Press, 2000); Fuller, *Thomas Kuhn.* Also see Geiger R, *Research and relevant knowledge: American research universities since World War II* (New Brunswick, NJ: Transaction, 2008); Wolfe AJ, *Competing with the Soviets: science, technology, and the state in Cold War America* (Baltimore: Johns Hopkins University Press, in press), chaps. 2 and 3.

27. Kleinman, *Politics;* Fuller, *Thomas Kuhn;* Geiger, What Happened after Sputnik?; Walsh J, NSF and its critics in congress: new pressure on peer review, *Science* 1975;188:999–1001; idem, Peer review—oops—merit review in for some changes at NSF, *Science* 1987;235:153; Agnew B, NIH eyes sweeping reform of peer review, *Science* 1999;286:1074–76; Kaiser J, NIH urged to focus on new ideas, new applicants, *Science* 2008;319:1169.

28. Forman P, Inventing the maser in postwar America, *Osiris* 1992;7:105–34; idem, Behind quantum electronics: national security as basis for physical research in the United States, 1940–1960, *Hist Stud Phys Sci* 1987;18:149–229; Statement by Wendell M. Stanley, February 27, 1950, Bancroft Manuscripts 78/18, Carton 21, Folder 46: Loyalty Oath, Online Archive of California (http://www.oac.cdlib.org/ark:/13030/hb3199p1nq/?brand=oac4); Anon., Faculty stirred by stand on oath, *NYT*, March 2, 1950:7 ("Mickey Mouse"); also Creager, *The life of a virus,* 261–63; Ninkovich FA, *The diplomacy of ideas: U.S. foreign policy and cultural relations, 1938–1950* (Cambridge: Cambridge University Press, 1981), conclusion; Osgood K, *Total cold war: Eisenhower's secret propaganda battle at home and abroad* (Lawrence: University of Kansas Press, 2006), 100 ("witch-hunters"). Hager T, *Force of nature: the life of Linus Pauling* (New York: Simon & Schuster, 1995); Badash L, The near-appointment of Linus Pauling at the University of California, Santa Barbara. In: *Physics in perspective* 2009;11:4–14; Wang J, Scientists and the problem of the public in cold war America, 1945–1960, *Osiris* 2002;17:323–47; Davenport D, Letters to FJ Allen: an informal portrait of Linus Pauling, *J Chem Education* 1996;73:21–28 ("weird insult").

29. Greenberg C, Avant-garde and kitsch. In: *Art and culture: critical essays* (Boston: Beacon Press, 1989), 3–21; Buck-Morss S, *Dreamworld and catastrophe: the passing of mass utopia in east and west* (Cambridge, MA: MIT Press, 2000):89. See also Ninkovich, *Diplomacy of Ideas;* Guilbaut S, *How New York stole the idea of modern art,* trans. Goldhammer A (Chicago: University of Chicago Press, 1983); Prevots N, *Dance for export: cultural diplomacy and the cold war* (Middletown, CT: Wesleyan University Press, 1998); Saunders FS,

Who paid the piper? the CIA and the cultural cold war (London: Granta, 1999), chap. 7; Osgood, *Total cold war*, chap. 7; Wilford H, *The mighty Wurlitzer: how the CIA played America* (Cambridge, MA: Harvard University Press, 2008), chap 5.

30. Needell AA, Truth is our weapon: project TROY, political warfare, and government-academic relations in the national security state, *Diplomatic History* 1993;17:399–420; Krige J, *American hegemony and the postwar reconstruction of science in Europe* (Cambridge, MA: MIT Press, 2006), chap. 2, chap. 7 et passim. Osgood, *Total cold war*, chap. 5. On Lysenkoism see deJong-Lambert W and Krementsov N, On Labels and issues: the Lysenko controversy and the cold war, *JHB* online first June 22, 2011 (DOI: 10.1007/s10739-011-9292-6); Selya R, Defending scientific freedom and democracy: the Genetics Society of America's response to Lysenko, *JHB* online first June 22, 2011 (DOI 10.1007/s10739-011-9288-2); Wolfe AJ, The cold war context of the golden jubilee, or, why we think of Mendel as the father of genetics, *JHB* online first June 22, 2011 (DOI 10.1007/s10739-011-9291-7). Anon, The secret of life, *Time* (Pacific Edition for South East Asia, Australasia, and Oceania), July 14, 1958, 34–38; also Rasmussen N, Midcentury biophysics bubble. Contrast this view of molecular genetics' Cold War political role to that of Kay (*Who wrote the book of life?*, chap. 3), who saw it mainly as a rhetorical prop for the cybernetic air defense system. In my view this rhetoric was of questionable impact on the American public and totally irrelevant to US allies outside NORAD's umbrella.

31. Geiger, *Research and relevant knowledge;* Strickland S, *Politics, science, and dread disease: a short history of United States medical research policy* (Cambridge, MA: Harvard University Press, 1972), and idem, *The story of the NIH grants programs* (Lanham, MD: University Press of America, 1988); Rasmussen N, Of small men, big science and bigger business: the Second World War and biomedical research in the United States, *Minerva* 2002;40:115–46; Poen M, *Harry S. Truman vs. the medical lobby: the genesis of Medicare* (Columbia MO: Missouri University Press, 1996); Gordon C, *Dead on arrival: the politics of health care in twentieth-century America* (Princeton: Princeton University Press, 2003). For a sense of the relative commitment to population health versus clinical and preclinical medical research, throughout the 1970s the NIEHS budget stood at about 2% and the NCI at about 30% of the total NIH budget (which itself accounted for ¾ of all Federal biomedical research; http://www.nih.gov/about/almanac/appropriations/index.htm), and US Department of Health and Human Services, NIH data book 1991, NIH Publication 91–126 (Washington, DC: NIH, 1991), tables 4 and 7. I thank William Summers for pointing out that generous NIH funding for medical schools also served Congressional factions seeking to subsidize and expand medical education, but this effect would have been concentrated in clinical fields and does not greatly affect the argument here concerning molecular biology.

32. Morange, *History*, chaps. 17–19; idem, The transformation of molecular biology; Kevles D, Renato Dulbecco and the new animal virology: medicine, methods, and, molecules, *JHB* 1993;26:409–42; Yi D, Cancer, viruses, and mass migration: Paul Berg's venture into eukaryotic biology and the advent of recombinant DNA research and technology, 1967–1980, *JHB* 2008;41:589–636.

33. Morange M, From the regulatory vision of cancer to the oncogene paradigm, 1975–1985, *JHB* 1997;30:1–29; idem, The transformation of molecular biology; idem, What history tells us. I. the operon model and its legacy, *J Biosci* 2005;30(3):313–16. I refer to the Nobel Prizes of 1975 (retroviruses), 1983 (transposition), 1987 (gene rearrangement in

antibody expression), 1989 (oncogenes), 1993 (transcript splicing), and 1995 (homeobox genes). Of molecular biology prizes in Physiology or Medicine in the interval only the 1978 prize for restriction enzymes and their use in mapping was not related specifically to issues of eukaryotic gene expression and evolution. See http://www.nobelprize.org/nobel_prizes/medicine/laureates.

34. On reverse transcriptase, see Baltimore, Discovery. A nice entrée into the dominant eukaryotic experimental systems for investigating regulation of gene expression in the 1970s is Breathnach R and Chambon P, Organization and expression of eucaryotic split genes coding for proteins, *Ann Rev Biochem* 1981;50:349–83. On sequencing see lectures of Maxam and Sanger at http://nobelprize.org/nobel_prizes/chemistry/laureates/1980.

35. Arber W, Host-controlled modification of bacteriophage, *Ann Rev Microbiol* 1965; 19:365–78; Smith HS and Wilcox KW, A restriction enzyme from *Hemophilus influenzae* I: purification and general properties, *JMB* 1970;51:379–91; Meselson M, Yuan R, and Heywood J, Restriction and modification of DNA, *Ann Rev Biochem* 1972;41:447–66.

36. Danna K, and Nathans D, Specific cleavage of simian virus 40 DNA by restriction endonuclease of *Hemophilus influenza*, *PNAS* 1971;68:2913–17; Roberts R, How restriction enzymes became the workhorses of molecular biology, *PNAS* 2005;102:5905–8. Rheinberger sees this recycling as "automatic" and technologically determined, but I disagree and find it dependent on institutional context and deliberate strategy; Rheinberger, *Toward a history*; Rasmussen, Mitochondrial structure.

37. Hall BD and Spiegelman S, Sequence complementarity of T2 DNA and T2 specific RNA, *PNAS* 1961;47:137–46; Temin H, Homology between RNA from Rous sarcoma virus and DNA from Rous virus-infected cells, *PNAS* 1964;52:323–29; Giacomoni D, The origin of DNA: RNA hybridization, *JHB* 1993;26:89–107; Marmur J, DNA strand separation, renaturation, and hybridization, *Trends Biochem Sci* 1994;19:343–46; Southern EM, Blotting at 25, *Trends Biochem Sci* 2000;25:585–88.

38. Roberts, How restriction enzymes; Southern, Blotting at 25; Sharp PA, Sugden B, and Sambrook J, Detection of two restriction endonuclease activities in *Haemophilus parainfluenzae* using analytical agarose-ethidium bromide electrophoresis, *Biochemistry* 1973;12:3055–63; Southern EM, Detection of specific sequences among DNA fragments separated by gel electrophoresis, *JMB* 1975;98:503–51.

39. Shapiro J, Machattie L, Eron L, et al., Isolation of pure *lac* operon DNA, *Nature* 1969; 224:768–74; Shapiro J, Eron L, and Beckwith J, *Nature* 1969;224:1337; Black H and Knox R, Research isolates a gene for 1st time, *Washington Post*, November 23, 1969:1; Reinhold R, Scientists isolate a gene; step in heredity control, *NYT*, November 23, 1969:1; Reinhold R, The gene: isolated-for good or evil? *NYT*, November 30, 1969:E10.

40. Morange, History, chap. 16, quote, 188; Jackson DA, Symons RH, and Berg P, Biochemical methods for inserting new genetic information into DNA of Simian Virus 40: circular SV40 DNA molecules containing lambda phage genes and the galactose operon of *Escherichia coli*, *PNAS* 1972;69:2904–9.

41. On the concept of the First and Second Cold Wars, see Halliday F, The making of the second cold war (London: Verso, 1983), chap. 1. For statistics on defense and other spending see Austin DA and Levit MR, Trends in Discretionary Spending, Congressional Research Service report RL34424, September 10, 2010 (available at fas.org). Wisnioski

M, Inside the system: engineers, scientists, and the boundaries of social protest in the long 1960s, *Hist and Technol* 2003;19:313–33; Wolfe, *Competing with the Soviets*, chap. 7.

42. Geiger, What happened after Sputnik, 358–59; Yi, Cancer, viruses. NIH and NSF budgets from http://www.nih.gov/about/almanac/appropriations/index.htm and http://dellweb.bfa.nsf.gov/NSFHist.htm respectively; inflation rates from http://www.usinflationcalculator.com/inflation/historical-inflation-rates/; NIH Ro1 and equivalent extramural grant application success rate data from http://www.nih.gov/UploadDocs/ Estimated_success_rates_1962–2008.xls.

43. Yi, Cancer, viruses; letter from Berg P, Baltimore D, Boyer HW, et al., *Science* 1974;185:303; Singer M and Sou D, *Science* 1973;181:1114; Hellman A, Oxman MN, and Pollack R, eds., *Biohazards in biological research* (Cold Spring Harbor: Cold Spring Harbor Laboratory, 1973).

44. Krimsky S, *Genetic alchemy: a social history of the recombinant DNA controversy* (Cambridge, MA: MIT Press, 1984), chap. 8; Wright S, *Molecular politics: developing American and British regulatory policy for genetic engineering, 1972–1982* (Chicago: University of Chicago Press, 1994), chap. 3; Reverby S, *Examining Tuskegee: the infamous syphilis study and its legacy* (Chapel Hill: University of North Carolina Press, 2009); Rothman DJ and Rothman SM, *The Willowbrook wars: bringing the mentally disabled into the community* (New Brunswick, NJ: Transaction, 2005); Fradkin PL, *Fallout: an American nuclear tragedy* (Tucson: University of Arizona Press, 1989); Hacker BC, Hotter than a $2 pistol: fallout, sheep, and the Atomic Energy Commission, 1953–1986, in Hevly B and Findlay JM, eds., *The atomic west* (Seattle: University of Washington Press, 1998); Daemmrich, A, A tale of two experts: thalidomide and political engagement in the United States and West Germany, *Soc Hist Med* 2002;15:137–58.

45. McElheny V, Gene transplants seen helping farmers and doctors, *NYT*, May 20, 1974:61; Morrow JF, Cohen SN, Chang ACY, et al., Replication and transcription of eukaryotic DNA in *Escherichia coli*, *PNAS* 1974;71:1743–47.

46. Yi, Cancer, viruses ("procedures" quote on 623); Hughes SS, Making dollars out of DNA: the first major patent in biotechnology and the commercialization of molecular biology, 1974–1980, *Isis* 2001;92:541–75 ("personal gain" paraphrase on 550). During academic life science's adjustment to the new era of business involvement, disapproval of entrepreneurial biologists was not uncommon; see Jones, M. Entrepreneurial science: the rules of the game. *SSS* 2009;39:821–51.

47. Krimsky, *Genetic alchemy*, 137–38; Wright, *Molecular politics*, chap. 3; Weiner C, Drawing the line in genetic engineering: self-regulation and public participation, *Persp Biology Med* 2001;44:208–20 (quoted is unnamed scientist at Asilomar, February 1975).

48. Krimsky, *Genetic alchemy*, chap. 14; Wright, *Molecular politics*, chap. 4; Gottweiss H, *Governing molecules: the discursive politics of genetic engineering in Europe and the United States* (Cambridge, MA: MIT Press 1998), chap. 3.

49. Angiotensin-2 and human insulin were mentioned by Boyer to Stanford intellectual property lawyer Niels Reimers; Hughes, Making dollars; Hughes-Swanson, 15–20; Metrick A and Yasuda A, *Venture capital and the finance of innovation* (New York: Wiley, 2011), chaps. 1, 7.

50. Chase M, Search for superbugs: industry sees a host of new products emerging from its growing research, *WSJ*, May 10, 1979, 48; Anon, FDA commissioner-to-be Ken-

nedy will move to end interagency communication gaps; private citizen Kennedy speaker at NAS forum on DNA, *FDC Reports,* March 14, 1977:3.

Chapter 2 · *The Insulin Trophy*

1. See the theoretical critique of this economic literature in Mirowski P, *Science mart* (Cambridge, MA: Harvard University Press, 2011). On organizational forms in biotech as predetermined, Powell WW and Sandholtz K, Amphibious entrepreneurs and the emergence of new organizational forms, *Strategic Entrepreneurship Journal* 2012;6:94–115. On the insulin supply, see Johnson IS, Potential benefits, in *Research with recombinant DNA: an academy forum, March 7–9, 1977* (Washington, DC: National Academy of Sciences, 1977):156–65; Anon, Lilly & Danish firm differ on possible shortage, *FDC Reports,* October 27, 1980:12; Anon, Nordisk pig pancreas insulin availability projections, *FDC Reports,* March 30, 1981:T&G 5–6. On humanization technology see Anon, Novo deriving human insulin from porcine pancreas: clinical trials in 1981, *FDC Reports,* September 22, 1980:12.

2. Hughes-Yansura, 18.

3. Shortt SED, Banting, insulin, and the question of simultaneous discovery, *Queen's Quarterly* 1982;89:260–73; Bliss M, *The discovery of insulin* (Chicago: University Chicago Press, 1982); Swann JP, *Academic scientists and the pharmaceutical industry* (Baltimore: Johns Hopkins University Press, 1988); Medvei VC, *A history of endocrinology* (Lancaster: MCP Press, 1982):454–70.

4. Vijayan M, The story of insulin crystallography, *Current Science* 2002;83:1598–1606; de Chadarevian S, Sequence, conformation, information: biochemists and molecular biologists in the 1950s, *JHB* 1996;29:361–86.

5. Baltimore D, Viruses, polymerases, and cancer, Nobel speech of December 12, 1975 (http://nobelprize.org/nobel_prizes/medicine/laureates/1975/baltimore-lecture.html); Temin H, The DNA provirus hypothesis, Nobel speech of December 12, 1975 (http://nobelprize.org/nobel_prizes/medicine/laureates/1975/temin-lecture.html); Morange M, From the regulatory vision of cancer to the oncogene paradigm, 1975–1985, *JHB* 1997; 30:1–29.

6. Rasmussen-Maniatis; Maniatis T, Cold Spring Harbor and recombinant DNA. In: Inglis JR, Sambrook J, and Witkowski J, eds., *Inspiring science: Jim Watson and the age of DNA* (Cold Spring Harbor: CSHL Press, 2003):321–27. In the piece describing parallel work with *human* globin, the investigators recount: "Some cDNA clones were initially obtained under Asilomar guidelines; work with these clones was stopped at the time of issuance of the draft of the NIH Guidelines and DNA from these cDNA plasmids was kept frozen until official adoption of the NIH Guidelines and certification of EK 2 host-vector systems. Cloning experiments were then resumed using X1776 and plasmid pCRl initially, then pMB9 after its official certification as an EK2 vector." Wilson JT, Wilson LB, deRiel JK, et al., Insertion of synthetic copies of human globin genes into bacterial plasmids, *Nucleic Acids Research* 1978;5:563–81.

7. Aviv H and Leder P, Purification of biological active globin messenger RNA by chromatography on oligothymidylic acid-cellulose, *PNAS* 1972;69:1408–12. The production of cDNAs from globin mRNA was simultaneously pursued by European rivals: Rougeon F and Mach B, Stepwise biosynthesis *in vitro* of globin genes from globin mRNA by DNA polymerase of avian myeloblastosis virus, *PNAS* 1976;73:3418–22.

8. Efstratiadis A, Kafatos F, Maxam AM, and Maniatis T, Enzymatic *in vitro* synthesis of globin genes, *Cell* 1976;7:279–88; Maniatis T, Kee SG, Efstratiadis A, and Kafatos F, Amplification and characterization of a globin gene synthesized *in vitro*, *Cell* 1976;8:163–82; Efstratiadis A, Kafatos F, and Maniatis T, The primary structure of rabbit globin mRNA as determined from cloned DNA, *Cell* 1977;10:571–85.

9. Hall S, *Invisible frontiers: the race to synthesize a human gene* (New York: Atlantic Press, 1987); Guarente L, Lauer G, Roberts TR, and Ptashne M, Improved methods for maximizing expression of a cloned gene: a bacterium that synthesizes rabbit ß-globin, *Cell* 1980;20:543–53.

10. Rasmussen-Lomedico; Hall, *Invisible frontiers*, chap. 8. Fuller F, A family of cloning vectors containing the *lacUV5* promoter, *Gene* 1982;19:43–54; Sgaramella V, Enzymatic oligomerization of bacteriophage P22 DNA and of linear simian virus 40 DNA, *PNAS* 1972;69:3389–93. For a related cloning strategy, see Guarente, Lauer, Roberts, and Ptashne, Improved methods. On the interpretation of Fuller's work outlined here, he would have used a vector such as pOP203–13, opened and digested judiciously with a nuclease like Bal 31 so as to expose the *lacZ* initiation codon at its 3′ end, joined by blunt-end ligation to a similarly exonuclease-trimmed cDNA beginning with the codon for the first amino acid in proinsulin. Screening would depend on insulin protein production, for instance with antibodies. In 1976 it was not certain that pre-proinsulin began with a methionine, ruling out the alternative strategy of employing the native start codon of pre-proinsulin; Chan SJ, Keim P, and Steiner DF, Cell-free synthesis of rat pre-proinsulins: characterization and partial amino acid sequence determination, *PNAS* 1976;73:1964–68.

11. Rasmussen-Villa-Komaroff; Hall, *Invisible frontiers*, chap. 8; Fuller, Family of cloning vectors; Villa-Komaroff L, Efstratiadis A, Broome S, et al., A bacterial clone synthesizing proinsulin, *PNAS* 1978;75:3727–31.

12. Rasmussen-Shine; Shine J and Dalgarno L, Determinants of cistron specificity in bacterial ribosomes, *Nature* 1975;254:34–38.

13. Hughes-Ullrich, 4–7. See Lucas-Lenard J, Protein biosynthesis, *Ann Rev Biochem* 1971;40:409–48.

14. Hughes-Ullrich, 7.

15. Hughes-Rutter, 108–16; Rasmussen-Lomedico; Hall, *Invisible frontiers*, chap. 3.

16. Hughes-Rutter, 108–16; *Invisible Frontiers*, chap. 7; Chirgwin J, Przybyla A, MacDonald R, and Rutter W, Isolation of biologically active ribonucleic acid from sources enriched in ribonuclease, *Biochemistry* 1979;18:5294–99.

17. Krimsky S, *Genetic alchemy: a social history of the recombinant DNA controversy* (Cambridge, MA: MIT Press, 1984), chap. 22; Hall, *Invisible frontiers*, chap. 4. For other useful accounts of the Cambridge affair, see Talbot B, Development of the National Institutes of health guidelines for recombinant DNA research, *Public Health Reports* 1983;98:361–68; Berg P and Singer M, The recombinant DNA controversy: twenty years later, *PNAS* 1995;92:9011–13; Wright S, Molecular biology or molecular politics? the production of scientific consensus on the hazards of recombinant DNA technology, *SSS* 1986;16:593–620; Weiner C, Drawing the line in genetic engineering: self-regulation and public participation, *Perspectives Biol Med* 2001;44:208–20.

18. Beckwith J, Recombinant DNA: does the fault lie within our genes? *Science for the People* 1977;9(3):14–17; Park B and Thatcher S, Dealing with the experts: the recombi-

nant DNA debate, *Science for the People* 1977;9(5):28–35; Walsh J, Science for the People: comes the evolution, *Science* 1976;191:1033–35; Wade N, Gene-splicing: critics of research get more brickbats than bouquets, *Science* 1977;195:466–69; Krimsky, *Genetic alchemy*, chaps. 22–23; Goodell R, Public involvement in the DNA controversy: the case of Cambridge, Massachusetts, *Science, Technology, & Human Values* 1979;4:36–43; Anon, DNA 'legislation is necessary,' PMA's Stetler, *FDC Reports*, March 21, 1977:T&G 3; Pfund N and Hofstadter L, Biomedical innovation and the press, *Journal of Communication* 1981;31:138–54.

19. See Frederickson D, *The recombinant DNA controversy: a memoir; science, politics, and the public interest, 1974–1981* (Washington, DC: American Society for Microbiology Press, 2001), chap. 7; Krimsky, *Genetic alchemy*, chap. 23; and also summary by Tooze J and Watson J, *The DNA story* (San Francisco: Freeman, 1981), 137–42 and chap. 6.

20. As Rifkin charged at the event (see below), a view essentially supported by Krimsky's contention that the meeting was called to head off local ordinances (*Genetic alchemy*, chap. 22), and also the "expert enclosure" interpretation advanced by Herbert Gottweiss in *Governing molecules: the discursive politics of genetic engineering in Europe and the United States* (Cambridge, MA: MIT Press, 1998).

21. The list of likely benefits offered up by the scientists included genetic engineering intervention in humans to counter genetic illness, and the promise of interferon through the drug industry, as well as basic knowledge of cancer and genetics; comments by Jonathan King, 38–41 (quote, 39). In: *Research with recombinant DNA: an Academy forum;* Nathan D, Potential benefits of the research, loc. cit., 49–54; Berg P, Potential benefits, 62–73 and comment on 110; King, 39.

22. Exchange between Stetten and Francine Simring, *Research with recombinant DNA*, 114–15. Anon, Private Citizen Kennedy Speaker at NAS Forum on DNA, *FDC Reports*, March 14, 1977:3. Lewin R, US genetic engineering in a tangled web, *New Scientist*, March 17, 1977:640–41. At the NAS symposium, human insulin not only served as the focus for pharmaceutical benefits and risks, debated by Ruth Hubbard and Irving Johnson, but also figured prominently in Daniel Nathan's antiregulation argument (Potential benefits). Gilbert waxed eloquently on the lifesaving value of human insulin at the July 7, 1976, meeting of the Cambridge City Council (Hall, *Invisible Frontiers*, 52–54). Rutter invoked a "cheap and limitless" supply of the lifesaving drug, superior to the animal hormone, when defending himself in a Senate Hearing (*Hearings before the Subcommittee on Science, Technology, and Space of the Committee on Commerce, Science, and Transportation, United States Senate*, Ninety-fifth Congress, first session, November 2, 8, and 10, 1977; Statement of William Rutter, 204).

23. Hughes-Ullrich, 9–10; Hughes-Rutter, 117, 141; Hall, *Invisible frontiers*, 95–99; Chirgwin, Przybyla, MacDonald, and Rutter, Isolation of biologically active ribonucleic acid. In the final report describing cloning into pMB9, the full-length pre-proinsulin cDNA was said to have been cloned with HindIII linkers and the two HaeIII subfragments with EcoRI linkers. Ullrich A, Shine J, Chirgwin J, et al., Rat insulin genes: construction of plasmids containing the coding sequences, *Science* 1977;196:1313–19.

24. Ullrich, Shine, Chirrwin, et al., Rat insulin genes; Chirgwin, Przybyla, MacDonald, and Rutter, Isolation of biologically active ribonucleic acid; Hall, *Invisible frontiers*, chaps. 8, 9; Scheller RH, Dickerson RE, Boyer H, et al., Chemical synthesis of restriction enzyme recognition sites useful for cloning, *Science* 1977;196:177–80.

25. Hughes-Rutter, 170–73. For registered letters and Goodman discussions with Genentech and Lilly representatives, see The Regents of the University of California v. Eli Lilly and Company, MDL Docket no. 912, IP-92-0224-C-D/G, United States District Court for the Southern District of Indiana, Indianapolis Division, 1995 US Dist. LEXIS 19003; 39 USP.Q.2D (BNA) 1225; December 11, 1995, by Judge S. Hugh Dillin [hereafter, Regents v. Lilly]. For further insight into why neither commercial relationship developed, see Hughes-Rutter, 175–76, where Rutter mentions that Goodman (with Ullrich and Shine) contracted with Genentech to convert the rat proinsulin clones to human through a semisynthetic approach, but that Rutter would not support this transfer of the university rights. Compare Ullrich-Hughes, 23, where Ullrich comments that Rutter spoiled a UCSF-Genentech collaboration deal by demanding almost as much equity as Goodman.

26. The Federal District Court decided that Ullrich's notebooks and other evidence argued against any recloning into pMB9 before the *Science* paper or the applications for patents describing that work, and that most of the insulin sequences described in *Science* and the patent applications could only have come from the original pBR322 clones, the DNA from which had never been destroyed; Regents v. Lilly, also Rasmussen-Gilbert. John Shine recalled more recently that he certainly did reclone the original insulin cDNA into pMB9 in April 1977, but wondered whether the cDNA might have been "spiked" with insert DNA from the original pBR322 clones (cf. chap. 3); Rasmussen-Shine; Cohn V, Scientists duplicate rat insulin gene: major DNA research breakthrough, *Washington Post*, May 24, 1977:A1, A12; Schmeck H Jr, Scientists report using bacteria to produce the gene for insulin: bacteria used to make insulin, *NYT*, May 24, 1977:73; Hughes-Ullrich, 20, and press conference cited there by Hughes: May 23, 1977, UCSF News Services, William J. Rutter correspondence, carton 6, folder: various press releases, UCSF archives.

27. Regents v. Lilly; Hughes-Rutter, 142, 173.

28. The postdocs would support the University's patent position in public and eventually in court. For example, Hall's *Invisible frontiers*, based on interviews with the postdocs before the litigation (but after the patents were revised to include them), supported the clone destruction and recloning version of events at UCSF later discredited by the District Court. On the patent filing, see *Invisible frontiers*, 142–43; Hughes-Ullrich, 20; Hughes-Rutter, 195. US Patents 4,440,859 and 4,652,525, pertaining to recombinant insulin genes and naming Ullrich as an inventor, refer to abandoned applications 801343, 805023, and 897709, of May 1977, June 1977, and April 1978, on which Ullrich may not have been named.

29. Bliss, *Discovery of insulin*, chap. 7; Hughes-Swanson, 17. Market size around 1981 was estimated at $150m in the US and $400m globally; Chase M, Genentech's insulin excites doctors less than it did brokers, *WSJ*, November 2, 1982:21.

30. Lilly's competitor Novo led in this technology but Lilly too was investigating it; Hughes-Johnson, 28; Anon, Novo deriving human insulin.

31. Hughes-Swanson, 21–22.

32. Mossman K, Profile of Arthur D Riggs, *PNAS* 2010;107:5269–71. Itakura K, Katagiri J, Narang S, et al., Chemical synthesis and sequence studies of deoxyribooligonucleotides which constitute the duplex sequence of the lactose operator of *Escherichia coli*, *JBC* 1975;250:4592–5000; Heyneker HL, Shine J, HM Goodman, et al., Synthetic *lac* operator DNA is functional *in vivo*, *Nature* 1976;263:748–52; Hughes-Heyneker, 37; Hughes-Swanson, 26–27; Hughes-Riggs, 34–35, 52–53.

33. Hughes-Riggs, 34–36, 39–41, 46; Hughes-Boyer, 75; Hughes-Itakura, 26; Hughes-Heyneker, 45–47; Itakura K, Hirose T, Crea R, et al., Expression in *Escherichia coli* of a chemically synthesized gene for the hormone somatostatin, *Science* 1977;198:1056–63; McElheny VK, Coast concern plans bacteria use for brain hormone and insulin, *NYT*, December 2, 1977:D1. Itakura K and Riggs R, Recombinant cloning vehicle microbial polypeptide expression, US Patent 4,704,362, issued November 3, 1987 on application of November 5, 1979 (continuing application first filed November 8, 1977).

34. Hall, *Invisible frontiers*, chap. 11.

35. Rasmussen-Villa Komaroff.

36. Wilson JT, Wilson LB, deRiel, JK, et al., Insertion of synthetic copies of human globin genes into bacterial plasmids, *Nucleic Acids Res* 1978;5:563–81; Gilbert W, Broome SA, Villa-Komaroff L, and Efstratiadis A, Protein synthesis, US Patent 4,411,994, issued October 25, 1983 on application of June 8, 1978; Anon, Harvard, Biogen in patent deal: gene splicing is licensed, *NYT*, November 29, 1983:D5.

37. Villa-Komaroff interview; Hall, *Invisible frontiers*, chap. 15; Villa-Komaroff L, Efstratiadis A, Broome SA, et al., A bacterial clone synthesizing proinsulin, *PNAS* 1978;75:3727–31.

38. Cordell B, Bell G, Tischer E, et al., Isolation and characterization of a cloned rat insulin gene, *Cell* 1979;18:533–43.

39. Hughes-Ullrich, 20–21; Hughes-Johnson, 30.

40. For example, US patent application 805023 was filed in June 1977, and 897709 in April 1978. On the delay in alerting the biosafety committee, see Hughes-Rutter, 141–42, 171–73, 180–81. Hopson J, Recombinant lab for DNA and my 95 days in it, *Smithsonian*, June 1977:54–62, quotes on 55, 60–61.

41. For "rumours" and the limited June disclosure of the incident to the Biosafety Committee, see Regents v. Lilly. Wade N, Recombinant DNA: NIH rules broken in insulin gene project, *Science* 1977;197:1342–45; Hughes-Rutter, 152–53, 181. DNA Kennedy bill will be withdrawn, *FDC Reports*, October 3, 1977: T&G 7–8; Frederickson, *Recombinant DNA controversy*, chap. 8.

42. *Regulation of recombinant DNA research: hearings before the subcommittee on science, technology, and space of the committee on commerce, science, and transportation, United States Senate*, Ninety-fifth Congress, first session, November 2, 8, and 10, 1977. Opening statements and questions by Senators Stevenson and Schmitt, 1–4 et passim, Statements and testimony of Dr. Philip Handler, 4–25, Statement of Dr. Paul Berg, 34–39, all on November 2, Frederickson testimony of November 8, 146–59, 175–86, at 184.

43. *Regulation of recombinant DNA research*, November 8, Senator Stevenson at 158; Ibid., Statement of Dr. William Rutter (accompanied by Boyer), 200–224 (especially 202); Hughes-Rutter, 181–83; Hughes-Boyer, 64–66.

44. *Regulation of recombinant DNA research*, November 8, queries of Sen Schmitt to Rutter, Boyer, 217–19 (patentability exchange on 217–18). In context it was possible to construe Rutter's statement that all "clones" had been destroyed meant only the transformed bacteria rather than cloned DNA too, and the statement that the pBR322 vector was not used after March meant that it was not used for further transformation (as opposed, for instance, to DNA sequencing). Of course these narrow construals were not necessarily the understandings of the Senators. Rutter maintained that to his knowledge all pBR322 "materials" were destroyed and that clerical error alone accounts for the inclusion of

pBR322 sequence data in the *Science* article describing insulin cloning in pMB9. For "not candid," see Regents v. Lilly. The appeals court decision that overturned the district court's inequitable conduct ruling on the enforceability of the UC patent did not touch these findings, but only the technical point of their materiality. Marshall E, A bitter battle over insulin gene, *Science* 1997;277:1028–30. For Boyer's feeling that the subcommittee was an Inquisition trying to "nail us to the cross," see Hughes-Boyer, 63–65; for Stevenson parting quote, Frederickson, *The recombinant DNA controversy*, 175. See also Wright S, *Molecular politics: developing American and British regulatory policy for genetic engineering, 1972–1982* (Chicago: University of Chicago Press, 1994), chap. 6.

45. McElheny, Coast concern. On the contract see Bader M, "Funds from medical institute: biochemistry split over Hughes issue," *UCSF Synapse* May 24, 1979:1, 4–5; McKelvey, *Evolutionary innovations*:131–32.

46. Hughes-Ullrich, 26–28; Hughes-Kleid, 10–16, 32–35, 39; Hughes-Goeddel, 17–18.

47. Hughes-Goeddel, 113; Hughes-Itakura, 47; Hughes-Kleid, 40–48; Crea R, Kraszewski A, Hirose T, and Itakura K, Chemical synthesis of genes for human insulin, *PNAS* 1978;75:5765–69; Goeddel DV, Kleid DG, Bolivar F, et al., Expression in *Escherichia coli* of chemically synthesized genes for human insulin, *PNAS* 1979;76:106–10.

48. Hughes-Kleid, 51, 54–55; Bronson G, Bacteria induced to produce insulin identical to humans': gene transplant techniques used to develop product that may be marketable, *WSJ*, September 7, 1978:17; Cohn V, Scientists in California create gene to make human insulin, *Washington Post*, September 7, 1978:1.

49. Bell G, Swain W, Pictet R, et al., Nucleotide sequence of a cDNA clone encoding human pre-proinsulin, *Nature* 1979; 282:525–27. Cases of crowds and markets determining scientific credit fit with Philip Mirowski's critique of "neoliberal science"; Lave R, Mirowski P, and Randalls, S, STS and neoliberal science, *SSS* 2010;40:659–75. I cannot agree with his supposition that such things were unprecedented in twentieth-century biology, even if the insulin story is such a case. See Rasmussen N, The moral economy of the drug company-medical scientist collaboration in interwar America, *SSS* 2004;34:161–86.

50. Hughes-Kleid, 55, 57–59, 85–87; Wetzel R, Kleid D, Crea R, et al., Expression in *Escherichia coli* of a chemically synthesized gene for a "mini-c" analog of human proinsulin, *Gene* 1981;16:63–71; Kleid D, Yansura D, Heyneker H, and Miozzari G, Bacterial polypeptide expression employing *tryptophan* promoter-operator; US Patent 5,888,808, issued March 30, 1999, on application of April 29, 1993 (continuing application first lodged March, 1980). Genentech applied to the NIH RAC to scale up recombinant production of proinsulin in early 1980, indicating that the new expression system was then substantially complete; Anon, Genentech preparing for large scale proinsulin production, *FDC Reports*, March 10, 1980:T&G 5–6. On protein renaturation see Builder SE and Ogez JR, Purification and activity assurance of precipitated heterologous proteins, US Patent 4,511,502, issued April 15, 1985 on application of June 1, 1984 (continuing applications first lodged December 1982).

51. Recombinant DNA Technology, *FDC Reports*, April 14, 1980:15; Anon, Recombinant DNA at FDA, *FDC Reports*, June 9, 1980:3–6 (quote on 3). On the fear of Japanese dominance, see for example, Anon, Who's ahead, who's behind, *Nature* 1980;283:123, an item cited in *amicus* briefs submitted to the Supreme Court in the Chakrabarty case (see below).

52. Anon, Genentech generates more operating income than any other gene-splicing firm, president Swanson maintains, *FDC Reports,* December 7, 1981:8; Russell C, FDA approves insulin made by splicing genes, *Washington Post,* October 30, 1982:A6; Chase M, Genentech's insulin excites doctors less than it did brokers, *WSJ,* November 2, 1982:21; Hughes-Johnson, 32.

53. Kevles D, Ananda Chakrabarty wins a patent: biotechnology, law, and society, 1972–1980, *Hist Stud Phys Biol Sci* 1994;25:111–35; Diamond v. Chakrabarty, 1980;447 US 303–22, 100, S. Ct.

54. Dickson D, Patenting living organisms—how to beat the bug-rustlers, *Nature* 1980;283:128–29; Anon, DNA on Wall Street, *FDC Reports,* June 9, 1980:T&G 8–9; Anon, Strong patent incentives needed to counter trade secrets trend, Wegner asserts, *FDC Reports,* July 28, 1980:4–5; Anon, DNA patent issue important but not crucial, Library of Congress, *FDC Reports,* November 10, 1980:T&G 5. In its Chakrabarty brief, the industry contended that "the absence of patents on living organisms will not preclude research or commercial exploitation in areas such as genetic engineering . . . and will only serve to limit public disclosure"; "Brief on Behalf of the Pharmaceutical Manufacturers Association, Amicus Curiae," with Diamond v. Chakrabarty, Docket no. 79–136, 447 US 303–22 (http://www.ipmall.info/hosted_resources/chisum_cases/briefs/15_diamond/15_diamond_4.htm).

55. "Brief of Dr. Leroy E. Hood, Dr. Thomas P. Maniatis, Dr. David S. Eisenberg, The American Society Of Biological Chemists, the Association of American Medical Colleges, the California Institute of Technology, the American Council on Education as *AMICI CURIAE,*" January 28, 1980, with Diamond v. Chakrabarty, Docket no. 79–136, 447 US 303–22 (http://www.ipmall.piercelaw.edu/hosted_resources/chisum_cases/briefs/15_Diamond/15_diamond_7.htm).

56. "Brief on Behalf of The American Society for Microbiology, Amicus Curiae," January 20, 1980, with Diamond v. Chakrabarty, Docket no. 79–136, 447 US 303–22 (http://ipmall.info/hosted_resources/chisum_cases/briefs/15_Diamond/15_diamond_10.htm); Diamond v. Chakrabarty decision.

57. Berman EP, Why did universities start patenting?: institution-building and the road to the Bayh-Dole act, *SSS* 2008;38:835–71.

58. Baltimore argued, against Krimsky, that a favourable "balance" between the needs of industrial sponsors should and probably would be obtainable in the emerging biotech "mixed economy," where academics "remain consultants and not turn their laboratories into factories for the solution of corporate problems." Baltimore D and Krimsky S, The ties that bind or benefit, *Nature* 1980;283:130–31 (on 130).

59. Krimsky, *Genetic alchemy,* chaps. 11, 22, 304 et passim; Wright, *Molecular Politics,* chap. 8; Anon, DNA committee has its critics, *Nature* 1975;257:637. Although I arrived at this analogy independently, I must grant priority to Duncan D, *The geneticist who played hoops with my DNA . . . and other masterminds from the frontiers of biotech* (New York: HarperCollins, 2005).

Chapter 3 · Growing Pains

1. Orsenigo L, *The emergence of biotechnology: institutions and markets in industrial innovation;* Chandler A Jr, *Shaping the industrial century: the remarkable story of the evolution of the modern chemical and pharmaceutical industries* (Cambridge, MA: Harvard

University Press, 2005); Vettel E, *Biotech: the countercultural origins of an industry* (Philadelphia: University of Pennsylvania Press, 2006; London: Pinter, 1989). For an equally dramatic but negative interpretation, see the key early statement of the position, Kenney M, *Biotechnology: the university-industrial complex* (New Haven: Yale University Press, 1988), and a recent addition to this interpretive tradition, Mirowski P, *Science mart: privatizing American science* (Cambridge, MA: Harvard University Press).

2. Boyer evidently considered growth hormone at an early stage because his son had been tested for deficiency (Hughes-Boyer, 42, 72). As we will see, Seeburg and Baxter began their pursuit of the hormone independently. Chase M, Search for superbugs: industry sees a host of new products emerging from its growing research, *WSJ*, May 10, 1979:48 ("same damn list"). The list of important publications on simultaneous discovery is too long to recite. For starting points see Kuhn T, Energy Conservation as an example of simultaneous discovery. In: Clagett M, ed., *Critical Problems in the history of science: proceedings of the Institute for the History of Science, 1957* (Ann Arbor: University of Michigan Press, 1959):321–56; Lakatos I, History of science and its rational reconstructions. In: Buck RC and Cohen RS, eds., *PSA 1970: in memory of Rudolf Carnap; Boston Studies in the Philosophy of Science VIII* (Dordrecht: Reidel, 1971):91–136.

3. Li CH and Evans HM, The isolation of pituitary growth hormone, *Science* 1944;99:183–84; Li CH, Evans HM, and Simpson ME, Crystallization of hypophyseal growth hormone, *Science* 1948;108:624–25; Hughes-Rutter, 43–46; Li CH, Growth hormone and recovery thereof. US Patent 3,118,815, issued January 21, 1964, on application of September 28, 1959; Li CH, Synthetic human pituitary growth hormone and method of producing it. US Patent 3,853,832, issued December 10, 1974, on application of April 27, 1971.

4. Bewley TA, Li CH, The chemistry of human pituitary growth hormone, *Adv Enzymol* 1975;42:73–166; Merimee TJ and Rabin D, A survey of growth hormone secretion and action, *Metabolism*, 1973;22:1235–51.

5. Lewis RA, Klein R, and Wilkins L, The effect of pituitary growth hormone in dwarfism with osseous retardation and hypoglycemia and in a cretin treated with thyroid, *J Clin Invest* 1950; 29:460–64; Matsuzaki F and Raben MS, Growth hormone, *Ann Rev Pharmacol* 1965;5:137–50; Niall HD, Hogan ML, Sauer R, et al., Sequences of pituitary and placental lactogenic and growth hormones: evolution from a primordial peptide by gene reduplication, *PNAS* 1971;68:866–69. Bewley TA, Dixon JS, and Li CH, Sequence comparison of human pituitary growth hormone, human chorionic somatomammotropin, and ovine pituitary growth and lactogenic hormones, *Int J Peptide Prot Res* 1972;4:281–87. McKelvey M, *Evolutionary innovations: the business of biotechnology* (Oxford: Oxford University Press, 2000):118 (quote) et passim.

6. Sutcliffe JG, pBR322 and the advent of rapid DNA sequencing, *Trends Biochem Sci* 1995;20:87–90.

7. Morange M, The transformation of molecular biology on contact with higher organisms, 1960–1980: from a molecular description to a molecular explanation, *Hist Phil Life Sci* 1997;19:369–93; idem, What history tells us: I, the operon model and its legacy, *J Biosci* 2005;30:313–16. I take the operon model, interpreted more broadly than Morange, to include the Britten-Davidson regulation model and therefore to be widely influential in eukaryotic molecular biology of the 1970s.

8. Rasmussen-Seeburg.

9. Rasmussen-Seeburg; Rasmussen-Shine; Stockton W, On the brink of altering life, *NYT* Sunday Magazine, February 17, 1980:4ff.

10. Rasmussen-Seeburg; Hughes-Rutter, 192

11. Rasmussen-Seeburg; Stockton, On the brink.

12. Rasmussen-Seeburg.

13. Martial JA, Baxter JD, Goodman HM, and Seeburg PH, Regulation of growth hormone messenger RNA by thyroid and glucocorticoid hormones, *PNAS* 1977;74:1816–20; Seeburg PH, Shine J, Martial JA, et al., Nucleotide sequence of part of the gene for human chorionic somatomammotropin: purification of DNA complementary to predominant mRNA species, *Cell* 1977;12:157–65. There was much technically clever about the UCSF team's strategy. For instance, both *Hha*I and *Hae*III cut single-as well as double-stranded DNA, which would certainly improve "clean-ness" of bands in a heterogeneous cDNA preparation. Also, to improve transformation rates of vectors with cDNA inserts, ligation vectors were stripped of terminal hydroxyl groups after opening with restriction enzymes so that they could not self-ligate, an innovation of Shine's.

14. Well characterized, 99% pure DNA fragments could be cloned under one level less stringent physical containment than otherwise, under early NIH guidelines.

15. Seeburg PH, Shine J, Martial JA, et al., Nucleotide sequence and amplification in bacteria of structural gene for rat growth hormone, *Nature* 1977;270:486–94.

16. Rasmussen-Seeburg; Shine J, Seeburg P, Martial JA, et al., Construction and analysis of recombinant DNA for human chorionic somatomammotropin, *Nature* 1977;270:494–99.

17. Rasmussen-Seeburg; Hughes-Rutter, 137, 196; Seeburg PH, Shine J, Martial JA, et al., Synthesis of growth hormone by bacteria, *Nature* 1978;276:795–98; Bader M, Funds from medical institute: biochemistry split over Hughes issue, *UCSF Synapse* May 24, 1979:1, 4.

18. Goodman HM, Shine J, and Seeburg PH, Recombinant DNA transfer vectors, US Patent 4,363,877, issued December 14, 1982, on application of April 19, 1978, continuing an application first filed September 23, 1977. See especially ex. 5 for human growth hormone cloning. Mention of this cloning "unpublished result" is made in Fiddes J, Seeburg PH, Denoto FM, et al., Structure of genes for human growth hormone and chorionic somatomammotropin, *PNAS* 1979;76:4294–98; *As You Like It*, Act 2, Scene 7.

19. Hughes-Ullrich, 20; Cohn V, Scientists duplicate rat insulin gene: major DNA research breakthrough, *Washington Post*, May 24, 1977:A1, A12.

20. Shorett P, Rabinow P, and Billings P, The changing norms of the life sciences, *NBT* 2003;21:123–25; Hughes-Ullrich, 20; Hughes-Rutter, 195, US Patents 4,440,859 and 4,652,525, pertaining to recombinant insulin genes and naming Ullrich as an inventor, refer to abandoned applications 801343 and 897709 of May 1977 and April 1978 respectively, on which Ullrich may not have been named. Similarly, US patents 4,363,877 and 4,407,948, naming Seeburg as an inventor and pertaining to growth hormone family genes, refer to abandoned applications 836218 and 897710 of September 1977 and April 1978 respectively, on which Seeburg may not have been named. Merton RK, Priorities in scientific discovery: a chapter in the sociology of science, *American Sociological Review* 1957;22:635–59; Long, PO, Invention, authorship, "intellectual property," and the origin of patents: notes toward a conceptual history, *Technology and Culture* 1991;32:846–84;

Biagioli, M, *Galileo's instruments of credit: telescopes, images, secrecy* (Chicago: University of Chicago Press, 2007).

21. Rasmussen-Seeburg.

22. Rasmussen-Seeburg; Bader, Funds from medical institute; McKelvey, *Evolutionary Innovations*, 131. Kleid reports that after UC resubmitted patent applications naming Ullrich and Seeburg among the inventors, the UCSF professors had convinced the postdocs to sign confidentiality and exclusive consulting agreements with Lilly, unbeknownst to Genentech; Kleid-Hughes, 71.

23. Swanson and Boyer visited Sweden for negotiations in December 1977 and soon after signed a letter of intent to produce human growth hormone for Kabi, then actually signed the final contract in August 1978. McKelvey, *Evolutionary innovations*, 114–16, 121, 134–41.

24. Hughes-Kleid, 70.

25. Hughes-Heyneker, 98–99; Rasmussen-Seeburg; Rasmussen-Shine.

26. Rasmussen-Seeburg. Stephen Hall, *Invisible frontiers: the race to synthesize a human gene* (New York: Atlantic Press, 1987):281–82, reports that this freezer lockout was verified by the reports of several other scientists, although exactly what Seeburg was working on at the time must rest on his own account.

27. Hughes-Ullrich, 23.

28. Rasmussen-Seeburg; Marshall E, Startling revelations in UC-Genentech battle, *Science* 1999;284:883–86; Seeburg testimony, April 20, 1999, Transcript of Proceedings, 1073–92, Regents v. Genentech, CA 90–2232, labeled Exhibit E and attached to Plaintiff's Memorandum of Law in Support of Bio-Technology General's Motion for Reconsideration, April 27, 1999, BTG v. Genentech files.

29. Hughes-Goeddel, 129–32; Goeddel DV, Heyneker HL, Hozumi T, et al., Direct expression in *Escherichia coli* of a DNA sequence coding for human growth hormone, *Nature* 1979;281:544–48. On the 1980 agreement see Hughes-Kiley, 42–43.

30. Seeburg, direct examination. Seeburg has later recounted (Rasmussen-Seeburg): "I mean all I could have done is just for instance put this fragment into the cDNA material from Kabi," or into the UCSF pituitary cDNA. Spiking HaeIII-digested pituitary cDNA with the 550 base HaeIII fragment liberated from its vector might in lab notebooks have given the appearance of a fresh cloning from pituitary RNA, and in later steps would have been indistinguishable. Seeburg also recalled that when he arrived at Genentech he already had a copy of the plasmid containing the 550 base HaeIII fragment of human growth hormone that he had cloned, implying another motive for the midnight raid. Feder BJ, Genentech agrees to settle patent dispute, *NYT*, November 17, 1999:C11; Hughes-Kleid, 78–79.

31. Goeddel, Heyneker, and Hozumi, Direct expression; Hughes-Goeddel, 132; here Goeddel recounts somewhat bitterly, after resolution of the growth hormone patent dispute against Genentech on Seeburg's testimony, that the Genentech scientists gave Seeburg last authorship because Heyneker and he felt sorry for Seeburg at the time.

32. Martial JA, Hallewell RA, Baxter JD, and Goodman HM, Human growth hormone: complementary DNA cloning and expression in bacteria, *Science* 1979;205:602–7. On Genentech's switch from the *lac* to the *trp* operon for expression, see Hughes-Kleid, 57–58, and McKelvey, *Evolutionary innovations:* 201–2, 218–19. The crucial City of Hope

patent exclusively licensed to Genentech was first filed in November 1977: Itakura K and Riggs AD, Recombinant cloning vehicle microbial polypeptide expression, US Patent 4,704,362, issued November 3, 1987, on application filed November 5, 1979, continuing application of November 8, 1977.

33. Hughes-Kiley, 81; Goeddel, Heyneker, and Hozumi, Direct expression:544.

34. Anon, Rival claims staked over gene-spliced growth hormone, *Medical World News*, August 1979:20–24; McKelvey, *Evolutionary Innovations*, 158–62; Andreopoulos S, Gene cloning by press conference, *NEJM* 1980;302:743–46. On Lubrizol, see Wade N, Recombinant DNA: warming up for big payoff, *Science* 1979;206:663–65, and Hughes-Swanson, 53.

35. Hughes-Kiley, 23–26; Christensen K, labs and ledgers: gene splicers develop a product; new breed of scientist-tycoons, *WSJ*, November 24, 1980:1; Elia C. Genentech Inc.'s initial offering will provide chance to invest in nascent genetic research, *WSJ*, August 27, 1980:37. On the Kabi contract renegotiation, see Hughes S, *Genentech: the beginnings of biotech* (Chicago: University of Chicago Press, 2011), chap. 5. Hughes also notes that by the end of 1979 Genentech's investor owners gave up trying to sell the firm to a major drug company and decided take it public as an exit strategy instead. This decision required Swanson to plan the firm's first product in 1980.

36. Mckelvey, *Evolutionary innovations*:139, 186–96; Hughes-Kleid, 85–87.

37. McKelvey, *Evolutionary innovations*:218–19; Hughes-Kleid, 85–87; Hughes-Young, 17–19; Gray GL and Heyneker HL, Secretion of correctly processed human growth hormone in *E. coli* and *Pseudomonas*, US Patent 4,755,465, issued on July 5, 1988, application of April 25, 1983; Chang CN, Gray GL, Heyneker HL, and Rey MW, Secretion of heterologous proteins, US Patent 4,963,495, issued October 16, 1990, on application of October 5, 1984; Bochner BR, Olson KC, and Pai RC, Periplasmic protein recovery, US Patent 4,680,262, issued July 14, 1987, on application of October 5, 1984.

38. McKelvey, *Evolutionary innovations*:220. Genentech was relying on Kabi for FDA preclinical data. See Stebbing N, Olson K, Lin N, et al., Biological comparison of natural and recombinant DNA-derived polypeptides, in Guerigian J, ed., *Insulin, growth hormone, and recombinant DNA technology* (New York: Raven, 1981):117–31. This Genentech study built on and cited bioassays developed by Kabi, and in the question session a Genentech author cited unpublished Kabi data on growth hormone pyrogenicity (Stebbing on 131).

39. Gertner JM, Genel M, Gianfredi SP, et al., Prospective clinical trial of human growth hormone in short children without growth hormone deficiency, *J Pediatr* 1984;104:172–76; Kaplan SL, Underwood LE, August GP, et al., Clinical studies with recombinant-DNA-derived methionyl human growth hormone in growth hormone deficient children. *Lancet* 1986;1(8483):697–700. Remaining allergic reactions, once *E. coli* membrane components were eliminated, were probably against incompletely renatured or misfolded growth hormone; Hughes-Kleid, 88; Frasier D, The not-so-good old days: working with pituitary growth hormone in North America, 1956–1985, *J Pediatr* 1997;131(Supp. 1, part 2):S1–S4; McKelvey, *Evolutionary innovations*, 217; Hughes-Kleid, 88.

40. McKelvey, *Evolutionary innovations*:242–49; Kolata G, New growth industry in human growth hormone?, *Science* 1986;234:22–24; Sell D, Growth hormone proves to be tempting for athletes, *Washington Post*, August 10, 1986:C14; Anon, Competition in recombinant hGH, *Bio/Technology* 1987;5, 311; Anon, Lilly and Genentech settle out of

court, *Pharmaletter*, January 16, 1985 (http://www.thepharmaletter.com/file/37857/lilly
-and-genentech-settle-out-of-court.html); Fisher LM, Drug makers will settle patent fight,
NYT, January 6, 1995:D3; Feder BJ, Genentech agrees to settle patent dispute, *NYT*,
November 17, 1999:C11.

41. Weiss R, Doctors and drug makers may be overpromoting a profitable hormone
that makes children taller, *Washington Post*, March 15, 1994:10; Day K, Genentech, nonprofit link studied: agencies probe whether foundation helped sales, *Washington Post*,
August 16, 1994:C1; Nordenberg T, Maker of growth hormone feels long arm of law,
FDA Consumer, Sept.–Oct. 1999:33; King R Jr, Medicine: Genentech set to end probe for
$50 million, *WSJ*, April 9, 1999:B1; Ratner M and Gura T, Off-label or off-limits?, *NBT*
2008;26:867–75.

42. Vettel, *Biotech;* New legislation to control & restructure pharmaceutical industry,
FDC-Reports July 30, 1973:3–9; Mintz M, Kennedy seeks reform in pharmaceutical industry, *Washington Post*, July 23, 1973:A3. FDA Commissioner Jere Goyan, publicly expressed
his overmedication view many times; Anon, US "overmedicated": drug chief, *Chicago Tribune*, November 7, 1979:10; Anon, Medicine: yellow light for tranquilizers, *Time*, Mon-
day, July 21, 1980:53. Anon, PMA meeting: new public mood against "overmedication,"
FDC Reports, April 14, 1980:9–11; Anon, FDA airing "overmedication" PSAs on TV, *FDC
Reports*, September 8, 1980:10–11.

43. Hughes-Rutter, 124.

44. Hughes-Swanson, 64; Hughes-Heyneker, 37–38, 75–77 et passim.

45. Bourdieu P, *Homo Academicus*, trans. Collier P (Stanford: Stanford University
Press, 1988). Although self-consciousness of the situation is not required for my argument, one does find it: "The sixties were good. We were convinced that we would find
a job at the end. It was definitely the years of innocence," reflected Heyneker (Hughes-
Heyneker, 12–13).

46. Hughes-Heyneker, 143; Hughes-Kleid, 60; Kenney, M, Biotechnology and the
creation of a new economic space; Fortun, M, Human Genome Project and the acceler-
ation of biotechnology. In: Thackray A, ed., *Private science: biotechnology and the rise of
the molecular sciences* (Philadelphia: University of Pennsylvania, 1998):131–43, 182–201
respectively.

47. Merton RK, *The sociology of science: theoretical and empirical investigations* (Chicago: University of Chicago Press, 1973). Joseph Fruton has suggested that the autocratic
style of laboratory leadership in biochemistry, while always common, has been particularly predominant in medical school settings. Certainly it is distinct from the more indi-
vidualist and iconoclastic tradition of molecular genetics and (other) biophysical fields.
Fruton J, *Contrasts in scientific style: research groups in the chemical and biochemical sciences*
(Philadelphia: American Philosophical Society, 1990), chap. 6.

48. Hughes-Ullrich, 24, 40; Rasmussen-Seeburg; also see Hughes-Heyneker, 39.
Compare the distribution of authorship in Gilbert's group; Hall, *Invisible frontiers*, chap. 15.

49. Yang-Feng TL, Francke U, Ullrich A, Gene for human insulin receptor: localization to site on chromosome 19 involved in pre-B-cell leukemia, *Science* 1985;228:728–31;
McGrath JP, Capon DJ, Goeddel DV, and Levinson AD, Comparative biochemical properties of normal and activated human ras p21 protein, *Nature* 1984;310:644–49; Hughes-
Goeddel, 46–47, 52; Hughes-Ullrich, 22 ("atmosphere"); Hughes-Kiley, 19, 76.

50. Kenney, *Biotechnology*, and idem, "Biotechnology and the Creation of a New Economic Space."

51. Heyneker HL, Shine J, HM Goodman, et al., Synthetic *lac* operator DNA is functional *in vivo, Nature* 1976;263:748–52.

52. Beckwith J, Recombinant DNA: does the fault lie within our genes?, *Science for the People*, May-June 1977:14–17. My point about "basic" biomedical research in the US being not completely divorced from application is broader and distinct from, but not incompatible with, Lily Kay's attribution of a eugenic agenda to the Rockefeller Foundation's 1930s–1950s support for molecular biology. See Kay L, Problematizing basic research in molecular biology, in Thackray ed., *Private science:* 20–38.

53. Yi D, Cancer, viruses, and mass migration: Paul Berg's venture into eukaryotic biology and the advent of recombinant DNA research and technology, 1967–1980, *JHB* 2008;41:589–636.

54. "In the United States, many industrial biotechnology developments rest on the broad base of knowledge generated by university research in the biological sciences. Such research has been funded largely by the National Institutes of Health (NIH) and other public health-oriented sponsors. As a consequence, the first areas of application of new biotechnology in the United States have been in the pharmaceutical field." US Congress Office of Technology Assessment, *Commercial biotechnology: an international analysis, OTA-BA-218* (Washington, DC: GPO, January 1984):119.

55. Hughes-Goeddel, 42. Although I cannot address the complex issue of quantifying the impact of the Bayh-Dole Act, my argument fits a recent critique suggesting that commercial incentives were already adequate to spur commercial biotechnology beforehand. See Mowery DC and Ziedonis AA, Academic patent quality and quantity before and after the Bayh-Dole act in the United States, *Research Policy* 2002;31:399–418; Rai AK and Eisenberg RS, Bayh-Dole reform and the progress of biomedicine, *Law and Contemporary Problems* 2003;66:289–314. On Chakrabarty, see chap. 2.

Chapter 4 · The Interferon Derby

1. Panem S, *The interferon crusade* (Washington, DC: Brookings Institution, 1984), chap. 2; Pieters T, *Interferon: the science and selling of a miracle drug* (London: Routledge, 2005), chaps. 4–5; Cantell K, *The story of interferon: the ups and downs in the life of a scientist* (Singapore: World Scientific, 1998), chap. 40.

2. Jaroslovsky R, Elusive quest: cancer research drive, begun with fanfare, hits disillusionment, *WSJ*, October 24, 1978:1, 22 (Kennedy quote). The National Cancer Institute, which had already committed to spending $1m on interferon for testing in 1975, spotlighted its interferon program by renaming it as the Biological Response Modifiers program in 1978, and expanding it with high funding priority (and a Congressional earmark in 1979). Panem, *Interferon crusade*, chap 3; Anon, Natural body substance: $2 million test on cancer retardant, *Chicago Tribune*, August 30, 1978:2; Hixson J, Interferon: the cancer drug we've ignored, *New York Magazine*, September 4, 1978:59–64; Schmeck H Jr, Interferon, virus foe, comes of age, *NYT*, December 26, 1978:C1; Pieters, *Interferon*, chaps. 6–7. Recombinant production of interferon was promised by molecular geneticists as early as 1977 at the National Academy forum, in biologists' amicus briefs in the Chakrabarty case, etc. (chap. 2).

3. Cantell, *Story*:66; Knight E Jr, Interferon: purification and initial characterization from human diploid cells, *PNAS* 1976;73:520–23; Rubinstein M, Rubinstein S, Familletti P, et al., Human leukocyte interferon: production, purification to homogeneity, and initial characterization, *PNAS* 1979;76:640–44; Stewart W, *The interferon system* (New York: Springer-Verlag, 1979); Levy W, Shively J, Rubinstein M, et al., Amino-terminal amino acid sequence of human leukocyte interferon, 1980;*PNAS* 77:102–4; Tyrell DAJ, Research on interferon: a review, *J Roy Soc Med* 1981;74:145–46; Gordon J and Minks M, The interferon renaissance: molecular aspects of induction and action, *Microbiological Reviews* 1981;45:244–66.

4. Jacobs P, Scientists study body's "shock troops," *Los Angeles Times*, June 12, 1978:C5; Hixson, Interferon: cancer drug we've ignored:64; Cantell, *Story,* 196. Multibillion sales figures were actually reached by another general-purpose recombinant cancer drug in the early 2000s, erythropoietin (chap. 5).

5. Hughes-Goeddel, 30–32; Rubinstein, Rubinstein, and Familletti, et al., Human leukocyte interferon.

6. Hughes-Kiley, 22, 50 (quote), 85–86; Hughes-Goeddel, 31–32; Koeffler HP and Golde DW, Acute myelogenous leukemia: a human cell line responsive to colony-stimulating activity, *Science* 1978;200:1153–54; Maeda S, Mccandliss R, Gross M, et al., Construction and identification of bacterial plasmids containing nucleotide sequence for human leukocyte interferon, *PNAS* 1980;77:7010–13; Moore v. Regents of the University of California, 51 Cal. 3d 120, 271 Cal, rptr. 146, 793 P.2d 479, cert. denied 499 US 936 (1991).

7. Maniatis T, Fritsch E, and Sambrook J, *Molecular cloning: a laboratory manual* (Cold Spring Harbor, NY: Cold Spring Harbor Laboratory Press, 1982):329–49. See also Goeddel DV, Yelverton E, Ullrich A, et al., Human leukocyte interferon produced by *E. coli* is biologically active, *Nature* 1980;287:411–16, taking into account Goeddel's statement that "lots of different [screening] methods"—all molecular—were tried and did not work before the synthetic oligonucleotide work described there. Hughes-Goeddel, 32 et passim; Hughes-Kleid, 120 ("brute force").

8. Hughes-Kleid, 119. Assuming 250 colony spots per Petri plate and replicate, a conservative estimate, 40 plates carried in total 10,000 transformants for hybridization in these screening experiments. The actual number screened was probably much higher.

9. Hughes-Goeddel, 35–36.

10. Maeda, McCandliss, and Gross, et al., Construction; Hughes-Kiley, 85–86; Hughes-Heyneker, 116–17.

11. With reasonably abundant proteins to which antibody was available, the specific mRNA could be isolated by using antibody to capture polysomes, mRNAs bound to ribosomes in the process of translation; Gelvin S, Heizmann P, and Howell SH, Identification and cloning of the chloroplast gene coding for the large subunit of ribulose-1,5-bisphosphate carboxylase from *Chlamydomonas reinhardi, PNAS* 1977;74:3193–97.

12. See also Goeddel, Yelverton, Ullrich, et al., Human leukocyte interferon produced by *E. coli.*

13. Colby C and Denney DW Jr, Interferon production. US Patent 4,262,090, issued April 14, 1981, on application of June 4, 1979; McCormick FP, Innis MA, Ringold GM, Expression of interferon genes in Chinese hamster ovary cells. US Patent 5,376,567,

issued December 27, 1994, on an application filed January 9, 1992, continuing claims in an application originally filed November 1, 1982; Paul Rabinow, *Making PCR: a story of biotechnology* (Chicago: University of Chicago Press, 1996), chap. 2.

14. Weissmann C, The cloning of interferon and other mistakes, *Interferon* 1981;3:101–34, on 117.

15. Curtis P, Mantei N, van den Berg J, et al., Presence of a putative 15S precursor to beta-globin mRNA but not to alpha-globin mRNA in Friend Cells, *PNAS* 1977;74:3184–88; Tilghman S, Curtis P, Tiemeier D, et al., The intervening sequence of a mouse beta-globin gene is transcribed within the 15S beta-globin mRNA precursor, *PNAS* 1978;75:1309–13.

16. Weissmann, Cloning of interferon: 101–2; Rasmussen-Nagata.

17. Weissmann, Cloning of interferon.

18. Weissmann, Cloning of interferon:105–6; Cantell, *Story*, chap. 19 et passim.

19. Weissmann, Cloning of interferon; Taniguchi T, Fujii-Kuriyama Y, and Muramatsu M, Molecular cloning of human interferon cDNA, *PNAS* 1980;77:4003–6.

20. Weissmann, Cloning of interferon. These postdocs were Lorraine Johnsrud from Gilbert's group, Shige Nagata, who had been with Weissmann for a year working on a bacteriophage project, and Tada Taniguchi, who had been working with Curtis on mouse interferon. Cantell, *Story*, 165–66, 169–70; Rasmussen-Nagata.

21. Weissmann, Cloning of interferon; Rasmussen-Nagata.

22. Weissmann, Cloning of interferon; Rasmussen-Nagata; Cantell, *Story*, 171–72; Rubinstein, Rubinstein, and Familletti, et al., Human leukocyte interferon; Taniguchi, Fujii-Kuriyama, and Muramatsu, Molecular cloning of human interferon cDNA; Taniguchi T, Ohno S, Fujii-Kuriyama Y, et al., The nucleotide sequence of human fibroblast interferon cDNA, *Gene* 1980;10:11–15.

23. Moreover, it had to be assumed that bacteria would not digest the foreign protein and that some would be excreted; that the presence of a prehormone "leader" would not prevent biological activity of the bacterially produced interferon; and that the absence of sugar groups found on the mature human protein would not prevent activity.

24. Weissmann, Cloning of interferon; Rasmussen-Nagata; Rasmussen-Hall; Nagata S, Taira H, Hall A, et al., Synthesis in *E. coli* of a polypeptide with human leukocyte interferon activity, *Nature* 1980;284:316–20.

25. Weissmann, Cloning of interferon; Andreopoulos S, Gene cloning by press conference, *NEJM* 1980;302:743–46; Wright S, Recombinant DNA technology an its social transformation, 1972–1982, *Osiris* 19;802:303–60; Anon, The big IF in cancer, *Time*, March 31 1980:40–46.

26. Nagata, Taira, and Hall, Synthesis in *E. coli*; Taniguchi T, Mantei N, Schwarzstein M, et al., Human leukocyte and fibroblast interferons are structurally related, *Nature* 1980;285:547–49; Streuli M, Nagata S, and Weissmann C, At least three human type alpha interferons: structure of alpha 2, *Science* 1980;209:1343–47; Nagata S, Mantei N, and Weissmann C, The structure of one of the eight or more distinct chromosomal genes for human interferon-alpha, *Nature* 1980;287:401–8.

27. Declaration of Alan Hall, July 24, 1990, AH-L5 and AH L6 plasmids, in Hall Exhibit 1; SN35 lac expression construct and B0595-B-SN206 construct labelled LAC-AUG in Declaration of Werner Boll, July 26 1990, Boll Exhibit 3; both attached to Amgen's Motion to Issue Letter of Request for International Judicial Assistance, Decem-

ber 22, 1997, Exhibits Vol. II of III, Tab B, Schering v. Amgen files. Pending, for example, was Goeddel DV and Heyneker H, Microbial expression of a gene for human growth hormone, US Patent 4,601,980, granted July 22, 1986, on application filed March 9, 1982, and continuing application first filed July 5, 1979.

28. Rasmussen-Hall; Rasmussen-Nagata; Allen G and Fantes KH, A family of structural genes for human lymphoblastoid (leukocyte-type) interferon, *Nature* 1980;287:408–11; Nagata, Mantei, and Weissmann, The structure of one of the eight; Goeddel et al., Human leukocyte interferon produced by *E. coli.*

29. Nagata, Structure of one of the eight; Goeddel DV, Leung DW, Dull TJ, et al., The structure of eight distinct cloned human leukocyte interferon cDNAs, *Nature* 1981;290:20–26; Goeddel DV and Pestka S, Microbial production of mature human leukocyte interferons, US Patent 6,610,830, issued August 26, 2003, on application of April 21 1981, continuing an abandoned application first filed July 1 1980; Goeddel DV, Hybrid human leukocyte interferons, US Patent 4,414,150, issued November 8, 1983, on application of February 23, 1981, continuing one filed November 10, 1980; Hughes-Goeddel, 38; Hughes-Kiley, 50; Gray PW, Leung DW, Pennica D, et al., Expression of human immune interferon cDNA in *E coli* and monkey cells, *Nature* 1982;295:503–8.

30. Mark DF and Creasey A, Multiclass hybrid interferons, US Patent 4,569,908, issued February 11, 1986, on application of February 3, 1983, continuing application first filed January 19, 1982; Werenne J, Technology report: the biology of the interferon system, *Bio/Technology* 1983;1:480–83; on Amgen's consensus interferon, see below. Hughes-Goeddel, 33; Hughes-Kiley, 86–87. In 1985 the two large drug firms extended and formalized their agreement; Anon, Hoffmann–La Roche and Schering-Plough agree to bury the hatchet on leukocyte interferon patent infringement, *BLR* 4:226. United States Patent and Trademark Office's Decision in Interference No. 101, 601, Appeals Board judges Ronald Smith, Mary Downey, and William Smith, December 15, 1995, Exhibit F, in Exhibits to Defendant Amgen's Initial Memorandum on Claim Construction of US Patent No. 4,530,901, Vol. II of II, filed 18 May 1998, Schering v. Amgen files.

31. Anon, Hoffmann–La Roche sues University of California for declaratory judgment, *BLR* 1982;1:3–6 (quote); Anon, Hoffmann–La Roche settles KG-1 cell line litigation with University of California, *BLR* 1982;1:186–87. Hughes-Rutter, 88–89; on gentleman's agreements and this case see Wade N, University and drug firm battle over billion-dollar gene, *Science* 1980;209:1492–94, and on the "particularism" of gift exchange, Jones, M. Entrepreneurial science: the rules of the game. *SSS* 2009;39:821–51.

32. Shapin S, *A social history of truth: civility and science in seventeenth-century England* (Chicago: University of Chicago Press, 1994); Biagioli M and Galison P, eds., *Scientific authorship: credit and intellectual property in science publication* (London and New York: Routledge, 2003); Wright S, *Molecular politics: developing American and British regulatory policy for genetic engineering, 1972–1982* (Chicago: University of Chicago Press, 1994):57–58. On recent financialization more generally see McKenzie D, *An engine, not a camera: how financial models shape markets* (Cambridge, MA: MIT Press, 2006), chap. 9.

33. Baron M, Making money work: genetics—cloning cash or parting fools and money?, *Los Angeles Times*, June 14, 1981:G1 ("best scientists"). Teitelman G, *Gene dreams: Wall Street, academia, and the rise of biotechnology* (New York: Basic Books, 1989); Klausner A, And then there were two, *Bio/Technology* 1985;3:605–12 ("stocks are sold").

34. Donis-Keller H, Gilbert group newsletter, *Biolabs Midnight Hustler*, undated

(October 1978?):8 ("clone shark"); publication gift of Lydia Villa-Komaroff, Appadurai A, eds., *The social life of things* (Chicago: University of Chicago Press, 1986); Smith CW, *Auctions: the social construction of value* (Berkeley: University of California Press, 1989). These biotech IPOs were inverted compared with art auctions also in the sense that, in Bourdieu's terms, the social capital of the eminent scientists affiliated with the firms was converted into economic capital—although I do not necessarily accept that these categories were diametrically opposite in 1980s America. Bourdieu P, *Distinction: a social critique of the judgement of taste,* Nice R, trans. (Cambridge, MA: Harvard University Press, 1987).

35. Schmeck H Jr, US to process 100 applications for patents on living organisms: basic to research, *NYT,* June 18, 1980:A22; Cohen SN and Boyer H, Process for producing biologically functional molecular chimeras, US Patent 4,237,224, issued December 2, 1980, on application of January 4, 1979; Sally Smith Hughes, Making dollars out of DNA: the first major patent in biotechnology and the commercialization of molecular biology, 1974–1980, *Isis* 2001;92:541–75; Jacobs P, "Glamour stock" could help cancer patients: medical breakthrough reported, *Los Angeles Times,* January 21, 1980:3 ("derby"); Gilbert quoted verbatim in Weissmann, Cloning of interferon:116. Whitefield D, DNA: 2 obscure firms see it as key to vast profits, *Los Angeles Times,* March 10, 1980:E1 (Cetus quote); Hughes-Kiley, 22; Hughes-Kleid, 119.

36. The *Nature* paper appeared in the October 2 issue, two weeks before the IPO—which was set in motion soon after Genentech's June–July 1980 patent application and announcement that it cloned interferon alpha. More pertinent to Genentech's original plans for publicizing the IPO was probably the paper's acceptance in August, enabling boasting and distribution of peer-reviewed preprints without severely transgressing prevailing scientific decorum. Blyth Eastman, Paine Webber, and Hambrecht & Qvist. *GENENTECH, 1,000,000 Shares of Common Stock.* Preliminary prospectus, dated August 10, 1980 (from S1 Registration Statement, obtained from SEC), 17–19; Hughes-Kiley, 81–82; Hughes-Swanson, 101–2.

37. Anon, Business ticker: Genentech shares to go for $35, *Chicago Tribune,* October 14, 1980:C1; Elia C, Genentech's final prospectus shows revenue comes primarily from health-care, *WSJ,* October 16, 1980, 55; Whitefield, DNA: 2 Obscure Firms; Parisi A, Technology—elixir for US industry, *NYT,* September 28, 1980:F1, 22; Silverlight J, Cashing in on DNA, *The Observer (UK),* February 10, 1980:3; Rowe JL Jr, Designer genes are snapped up, *Washington Post,* October 15, 1980:E1; Anon, Shaping life in the lab and profiting from gene splicing, *Time,* March 9, 1981:36–49.

38. Storch C, Genentech hits $71 on 1st day, *Chicago Tribune,* October 15, 1980:C1 ("first chance"); Anon, Hot reception seen today for Genentech as first gene-splicing firm to go public, *WSJ,* October 14, 1980:6; Metz T, New genentech issue trades wildly as investors seek latest high-flier, *WSJ,* October 15, 1980:31; Rowe, JL Jr, Speculation fever seeping through Wall Street, *Washington Post,* November 2, 1980:G1; Elia C, Commercial potential of genetic engineering seems to be vast in coming decade, study says, *WSJ,* January 16, 1981:37; Wade, University and drug firm battle; Teitelman, *Gene Dreams,* 11–13 (and "stockbrokers" on 27).

39. Whitefield, DNA: 2 Obscure Firms; Anon, Cetus is second firm in the genetics field planning to go public, *WSJ,* December 3, 1980:20; Zonana V, Cetus, a genetic engineering firm, plans initial public offer of 5.2 million shares, *WSJ,* January 14, 1981, 37;

Dorfman D, Cetus: hottest '81 stock or a ripoff?, *Chicago Tribune*, February 8, 1981: N1; Lehman Brothers, Kuhn, Loeb and L.F. Rothschild, Unterberg, Towbin, *5,201,685 Shares Cetus Common Stock*, Preliminary prospectus, dated February 1981 (from S1 Registration Statement, obtained from SEC), 17–19.

40. Storch C, $120 million offering: Cetus debut "lackluster," *Chicago Tribune*, March 7, 1981:A6; Lueck T, Cetus in record offering; market response is cool: $119.6 million raised, *NYT*, March 7, 1981:31; idem, Cetus charting a broad course, *NYT*, June 5, 1981:D1 (quote); Anders G, Many new stock issues are being canceled, delayed or reduced by nervous underwriters, *WSJ*, October 7, 1981:55; Anon, Bristol-Myers buys Genex process rights for making interferon, *WSJ*, June 6, 1980:26; Anon, Genex Corp. files initial public offer, *WSJ*, September 2, 1982:24; Anon, Two million Genex shares are sold by First Boston, *WSJ*, September 30, 1982:44.

41. Anon, Biotechnology retrenchment: business now termed less profitable, *NYT*, February 19, 1983:29 ("impressive talent"); Anon, Biogen phase 1 clinicals on human immune interferon could begin in 1983 in Europe, *FDC Reports*, February 14, 1983:8–9; Anon, Biogen N.V. up to 23 1/4 in first day of trading, *NYT*, March 23, 1983:D11.

42. Stabinsky Y, Consensus human leukocyte interferon, US Patent 4,695,623, issued September 22, 1987, on application of December 12, 1983; Rathmann G, Biotechnology startups, in Moses V and Cape R, eds., *Biotechnology the science and the business* (London: Harwood, 1991):49–60; Anon, Amgen files for initial offering of 2 million common shares, *WSJ*, May 10, 1983:46; Anon, Briefs, *NYT*, June 14, 1983:D7; Sherwood EB, Lag in heady new issues hints of market weakness, *Christian Science Monitor*, July 25, 1983, 10; Rathmann-Hughes, 36 ("ungrateful"); Smith Barney, Harris Upham & Co and Dean Witter Reynolds and Montgomery Securities, *2,000,000 Shares Amgen Common Stock*, Preliminary prospectus, dated May 9, 1983 (from S1 Registration Statement, obtained from SEC), 4, 13–14; Metz R, Wall Street: biotech: high risk, high promise, *Los Angeles Times*, October 31, 1983:E2. On results of Chiron offering see US Congress Office of Technology Assessment, *Commercial biotechnology: an international analysis*, OTA-BA-218 (Washington, DC: GPO, January 1984):282–83. Gibbons A, Chiron buys Cetus: a tale of two companies, *Science* 1991;253:503. On the unsuccessful Cetus interleukin-2 trials, see Löwy I, *Between bench and bedside: science, healing, and interleukin-2 in a cancer ward* (Cambridge, MA: Harvard University Press, 1996). On Chiron's hepatitis B success see Valenzuela P, Gray P, Quiroga M, et al., Nucleotide sequence of the gene coding for the major protein of hepatitis B virus surface antigen, *Nature* 1979;280:815–9; Edman JC, Hallewell RA, Valenzuela P, et al., Synthesis of hepatitis B surface and core antigens in E. coli, *Nature* 1981;291:503–6; Davidson J, Lab-made vaccine for hepatitis b is cleared by FDA, *WSJ*, July 24, 1986:16.

43. Waldholz M, Hyped drug: ballyhoo has faded, but interferon still has boosters at high levels, Schering-Plough, *WSJ*, September 30, 1983:1, 16; Powledge TM, Interferon on trial, *Bio/Technology*, March 1984;2:214–28; Korzeniowsky C, Interferon break-through attracts even more money, *The Observer (UK)*, March 1, 1981:18.

44. Vecchio-Good M-J, Good B, Schaffer C, and Lind S, American oncology and the discourse on hope, *Culture, Medicine and Psychiatry* 1990;14:59–79. Anon, Shaping life in the lab; Anon, The big IF in cancer; Glaser VP, Gutterman talks about the trials, *Bio/Technology* 1984;2:227; Borden EC, Holland JF, Dao TL, et al., Leukocyte-derived inter-

feron (alpha) in human breast carcinoma, The American Cancer Society phase II trial, *Ann Intern Med* 1982;97:1–6; Quesada JR, Swanson DA, Trindade A, et al., Renal cell carcinoma: antitumor effects of leukocyte interferon, *Cancer Res* 1983;43:940–47.

45. Rasmussen N, The drug industry and clinical research in interwar America: three types of physician collaborator, *Bull Hist Med* 2005;79:50–80. For early, influential sources on sponsorship and other sources of bias, see Dickersin K, Chan S, Chalmers TC, Publication bias and clinical trials. *Control Clin Trials*, 1987;8:343–53; Chalmers I, Misconduct in medical research, *Brit Med J* 1989;298:256; Dickersin K, The existence of publication bias and risk factors for its occurrence, *JAMA* 1990;263:1385–89. For a recent review see Lundh A, Sismondo S, Lexchin J, et al., Industry sponsorship and research outcome, *Cochrane Database of Systematic Reviews* 2012, art. MR000033. In published randomised trials sponsorship predicts favourable publications by an odds ratio around 2 in oncology, an order of magnitude less than the bias found in some other specialities; Peppercorn J, Blood E, Winer E, Partridge A, Association between pharmaceutical involvement and outcomes in breast cancer clinical trials, *Cancer* 2007;109:1239–46; Booth CM, Cescon DW, Wang L, et al., Evolution of the randomized controlled trial in oncology over three decades, *J Clin Oncol* 2008;26:5458–64. This ratio is twice what it should be, given that the actual long-term success rate in phase III cancer trials of the highest quality (NCI sponsored trials) is about 50%; see Djulbegovic B, Kumar A, Soares HP, et al., Treatment success in cancer: new cancer treatment successes identified in phase III randomized controlled trials conducted by the National Cancer Institute-sponsored cooperative oncology groups, 1955 to 2006, *Arch Intern Med* 2008;168:632–42.

46. Henderson N, Cancer drug interferon wins approval for commercial use: genetically engineered cancer drug, *Washington Post*, June 5, 1986:A1; Glaspy JA, Jacobs AD, Golde DW, The UCLA experience with type I interferons in hairy cell leukemia, *Leukemia* 1987;1:323–26; Quesada JR, Reuben J, Manning JT, et al., Alpha interferon for induction of remission in hairy-cell leukemia, *NEJM* 1984;310:15–18; Quesada JR, Hersh EM, Manning JT, et al., Treatment of hairy cell leukemia with recombinant alpha-interferon, *Blood* 1986;68:493–97; Gutterman JU, Cytokine therapeutics: lessons from interferon alpha, *PNAS* 1994;91:1198–205. For oncologists' thinking about hairy cell and interferon alpha in 1989 see Murren JR and Buzaid AC, The role of interferons in the treatment of malignant neoplasms, *Yale J Biol Med* 1989;62:271–90.

47. Jacobs AD, Champlin RE, and Golde D, Recombinant a-2-Interferon for hairy cell leukemia, *Blood* 1985;65:1017–20; Golomb HM, Jacobs A, Fefer A, et al., Alpha-2 interferon therapy of hairy-cell leukemia: a multicenter study of 64 patients, *J Clin Oncol* 1986;4:900–905; Glaspy, Jacobs, Golde, et al., UCLA experience.

48. Henderson, Cancer drug interferon wins approval; Stipp D, FDA's approval of interferon to treat cancer boosts biotechnology concerns, *WSJ*, June 5, 1986:1; Anon, A groundswell of drugs from the biotech pipeline, *Chemical Week*, January 21, 1987:6; Anon, Fast approval for breakthrough drugs, *WSJ*, March 19, 1987:35; Cheson BD and Martin A, Clinical trials in hairy cell leukemia: current status and future direction, *Ann Intern Med* 1987;106:871–78. The 1997 FDA label for Intron A describes a difference between 40-month survival in a matched historical control group (http://www.accessdata.fda.gov/drugsatfda_docs/label/1997/ifnasch110697-lab.pdf), while the 2002 FDA label for Roferon-A refers to a study comparing the 2-year survival rate for Roferon patients to a matched historical control group (http://www.accessdata.fda.gov/drugsatfda_docs/

label/2002/ifnahof081502LB.pdf). On the discourse of "drug lag" see: Carpenter D, *Reputation and power* (Princeton: Princeton University Press, 2010):374–80; Tobbell D, *Pills, power, and policy: the struggle for drug reform in cold war America and its consequences* (Berkeley: University of California Press, 2012):183–87; Nik-Kah, E, Neoliberal pharmaceutical science and the Chicago school of economics, *SSS*, in press.

49. Dagher R, Johnson J, Williams G, et al., Accelerated approval of oncology products: a decade of experience, *JNCI* 2004;96:1500–1509; Djulbegovic, Treatment success in cancer. On the industry role in accelerated approval, see Davis C and Abraham J, Desperately seeking cancer drugs: explaining the emergence and outcomes of accelerated pharmaceutical regulation, *Sociology of Health & Illness* 2011;33:731–47.

50. Quesada JR, Hawkins M, Horning S, et al., Collaborative phase I-II study of recombinant DNA-produced leukocyte interferon (clone A) in metastatic breast cancer, malignant lymphoma, and multiple myeloma, *Am J Med.* 1984;77:427–32; Sherwin SA, Mayer D, Ochs JJ, et al., Recombinant leukocyte A interferon in advanced breast cancer: results of a phase 2 efficacy trial, *Ann Intern Med* 1983;98:598–602; Padmanabhan N, Balkwill FR, Bodmer JG, et al., Recombinant DNA human interferon alpha 2 in advanced breast cancer: a phase 2 trial, *Br J Cancer* 1985;5155–60; Fentiman IS, Balkwill FR, Cuzick J, et al., A trial of human alpha interferon as an adjuvant agent in breast cancer after loco-regional recurrence, *Eur J Surg Oncol* 1987;13:425–28.

51. Macheledt JE, Buzdar AU, Hortobagyi GN, et al., Phase 2 evaluation of interferon added to tamoxifen in the treatment of metastatic breast cancer, *Breast Cancer Res Treat* 1991;18:165–70; Buzdar AU, Hortobagyi GN, Kau SW, Adjuvant therapy with escalating doses of doxorubicin and cyclophosphamide with or without leukocyte alpha-interferon for stage II or III breast cancer, *J Clin Oncol* 1992;10:1540–46.

52. Kelly K, Crowley JJ, Bunn PA, et al., Role of recombinant interferon alfa-2a maintenance in patients with limited stage small cell lung cancer responding to concurrent chemoradiation: a Southwest Oncology Group study, *J Clin Oncol* 1995;13:2924–30. Ruotsalainen TM, Halme M, Tamminen K, et al., Concomitant chemotherapy and IFN-a for small cell lung cancer: a randomized multicenter phase 3 study, *J Interferon Cytokine Res* 1999;19:253–59; Bowman A, Fergusson RJ, Allan SG, et al., Potentiation of cisplatin by alpha-interferon in advanced non-small cell lung cancer (NSCLC): a phase II study, *Ann Oncol* 1990;1:351–53. Kohne-Wompner CH, Koschel G, Pawel G, et al., Carboplatin, etoposide, vincristine hemotherapy plus radiotherapy in limited disease small cell lung cancer, followed by maintenance with interferon-alpha 2b versus observation, *Ann Oncol* 1992;3(Supp. 5):40.

53. Edsmyr F, Esposti PL, Andersson L, et al., Interferon therapy in disseminated renal cell carcinoma, *Radiotherapy & Oncology* 1985;4:21–26; Hrushesky WJ and Murphy GP, Current status of the therapy of advanced renal carcinoma, *J Surg Oncol* 1977;9:277–88; Harris DT, Hormonal therapy and chemotherapy of renal-cell carcinoma, *Semin Oncol*, 1983;10:422–30; Quesada JR, Rios A, Swanson D, et al., Antitumor activity of recombinant-derived interferon a in metastatic renal cell carcinoma, *J Clin Oncol* 1985;3:1522–28; Quesada JR, Evans L, Saks SR, et al., Recombinant interferon alpha and gamma in combination as treatment for metastatic renal cell carcinoma, *J Biol Response Mod* 1988;7:234–39.

54. Dexeus FH, Logothetis CJ, Sella A, and Finn L, Interferon alternating with chemotherapy for patients with metastatic renal cell carcinoma, *Am J Clin Oncol* 1989;12:350–

54; Steineck G, Strander H, Carbin BE, et al., Recombinant leukocyte interferon alpha-2a and medroxyprogesterone in advanced renal cell carcinoma: A randomized trial, *Acta Oncologica* 1990;29:155–62.

55. Bergerat JP, Herbrech R, Dufour P, et al., Combination of recombinant alpha interferona-2a and vinblastine in advanced renal cell cancer, *Cancer* (Phila.) 1988;62:2320–24; Foon K, Doroshow J, Bonnem E, et al., A prospective randomized trial of alpha 2b-interferon/gamma-interferon or the combination in advanced metastatic renal cell carcinoma, *J Biol Response Modifiers* 1988;7:540–45; Schornagel JH, Verweij J, ten Bokkel Huinink WW, et al., Phase II study of recombinant interferon alpha-2a and vinblastine in advanced renal cell carcinoma, *J Urol* 1989;142:253–56; Fossa SD, Martinelli G, Otto U, et al., Recombinant interferon alfa-2a with or without vinblastine in metastatic renal cell carcinoma: results of a European multi-center phase 3 study, *Ann Oncol* 1992;3:301–5. The typically optimistic conclusion of this last study suggests that survival rates among the moderately sick patient subgroups appear higher than historical comparators.

56. Epstein S, Activism, drug regulation, and the politics of therapeutic evaluation in the AIDS era, *SSS* 1997;27:691–726; Goodman and Walsh V, *The story of taxol: nature and politics in the pursuit of an anti-cancer drug* (Cambridge: Cambridge University Press, 2001); Dagher, Johnson, Williams, et al., Accelerated approval.

57. Murren and Buzaid, The role of interferons; a critique of this logic can be found in Horoszewicz JS and Murphy GP, An assessment of the current use of human interferons in therapy of urological cancers, *J Urol* 1989;142:1173–80. Anon, Schering files suit to block interferon patent infringement, *BLR* 1989;8:183–84; Cureton G, Modern biotechnology: promise and performance, *Medical Marketing and Media,* June 1991, 56–73; Fischer L, Return of an ex-wonder drug: a former wonder drug is finding favor again, *NYT,* August 16, 1990:D1, D6 (quote, D6). Vecchio-Good, Good, Schaffer, and Lind, American oncology; Pieters T, Marketing medicines through randomised controlled trials: the case of interferon, *Brit Med J* 1998; 317:1231–33. Schering eventually pled guilty and paid fines for off-label marketing of Intron-A in cancer; Westphal SP, Seward ZM, and Carreyrou J, Schering-Plough settles charges for $435 million, *WSJ,* August 30, 2006:A2; McLaughlin P, US pharma company fined $435 million, *Can Med Assoc J* 2006;9:1046. For a history of clinical trials in oncology, albeit one failing to attend much to commercial sponsorship and its effects, see Keating P and Cambrosio A, *Cancer on trial: oncology as a new style of practice* (Chicago: University of Chicago Press, 2011).

58. Oliver RTD, Are cytokine responses in renal cell cancer the product of placebo effect of treatment or true biotherapy?, *Brit J Cancer* 1998;77:1318–20; Medical Research Council Renal Cancer Collaborators, Interferon-alpha and survival in metastatic renal cell carcinoma: early results of a randomized controlled trial, *Lancet* 1999;353:14–17; Coppin C, Porzsolt F, Kumpf J, et al., Immunotherapy for advanced renal cell cancer, *Cochrane Database of Systematic Reviews* 2000, issue 2:00142.

59. Aggarwal S, What's fueling the biotech engine?, *NBT* 2007;25:1097–1104; Brok J, Gluud LL, Gluud C, Ribavirin plus interferon versus interferon for chronic hepatitis C, *Cochrane Database of Systematic Reviews* 2010, issue 1:005445; Lee LY, Tong CY, Wong T, et al., New therapies for chronic hepatitis c infection: a systematic review of evidence from clinical trials, *Int J Clin Pract* 2012;66:342–55.

60. Here I deliberately invoke the concept of financial manias and investment bubbles; Mackay C, *Extraordinary popular delusions and the madness of crowds* (New York:

Wiley, 1996). Panem, *Interferon crusade*, chap. 3; Anon, Shaping life in the lab (Boyer on 39); Wade, University and drug firm battle ("rupture" on 1494).

Chapter 5 · Epo

1. Erslev A, The discovery of erythropoietin, *ASAIO J.* 1993;39:89–92.

2. Jacobson LO, Goldwasser E, Fried W, and Plzak L, Role of the kidney in erythropoiesis, *Nature* 1957;179:633–34; Goldwasser E and Kung C, Purification of erythropoietin, *PNAS* 1971;68:697–98; Goldwasser E and Kung C, The molecular weight of sheep plasma erythropoietin, *JBC* 1972;247:5159–60; Miyake T, Kung C, and Goldwasser E, Purification of human erythropoietin, *JBC* 1977;252:5558–64; Goldwasser E, Erythropoietin: a somewhat personal history, *Perspectives Biol Med* 1996;40:18–32.

3. Miyake et al., Purification; Goldwasser, Somewhat personal history; Transcript of deposition of Dr. Eugene Goldwasser, April 17–21, 1989, in Designated Transcript and Summaries Vol. 2:59, 63–64; Testimony of Thomas Wall, October 19, 1989, Transcripts, Vol. 37:31–34, 37–40; Alan S Levine (NHLBI), December 22, 1983, Erythropoietin distribution data June 1978 to present, Plaintiff's Exhibit 344, attached to Defendants' Joint Appendices to Proposed Findings of Fact and Conclusions of Law, Appendices, all in Amgen v. Chugai files.

4. Goldwasser, Somewhat personal history; the meeting must have been the Second Conference on Hemoglobin Switching, published as Stamatoyannopoulos G and Nienhuis AW, eds., *Organization and expression of globin genes* (Liss: New York, 1981); Goldwasser deposition, April 1989:146–54; Rathmann-Hughes, 29 (quote); Rathmann G, Biotechnology startups, in Moses V and Cape R, eds., *Biotechnology, the science and the business* (London: Harwood, 1991):49–60.

5. Goldwasser deposition April 1989:161, 167–69, 657–59; Goldwasser stockbroker statement, January 31, 1989, Defendant's exhibit 613, attached to Defendants' Joint Appendices to Proposed Findings of Fact and Conclusions of Law, Vol 3, II–96; both in Amgen v. Chugai files. Goldwasser deposition April 1989:161, 167–69.

6. Hewick RM, Hunkapiller MW, Hood LE, and Dreyer WJ, A gas-liquid solid phase peptide and protein sequenator, *JBC* 1981;256:7990–93; Testimony of Rodney Hewick, August 23, 1989, Transcripts Vol. 11:30–32; sequence entry in Hewick Lab notebook of March 1984, page 20, Plaintiff's Exhibit 588, attached to Defendants' Joint Appendices to Proposed Findings of Fact and Conclusions of Law, Vol. 7, XI–2; both in Amgen v. Chugai files. On Golde as source, see Memorandum and Order of Magistrate Patti Saris, December 11, 1989, *BLR* 1990;9:25–71, on 60.

7. Goldwasser to Mach, October 9, 1981, Defendant's Exhibit 390, attached to Defendants' Joint Appendices to Proposed Findings of Fact and Conclusions of Law, Vol. 3, II–101; Goldwasser deposition, April 1989:47, 152–60; Testimony of Dr. Julian Davies, August 17, 1989, Transcripts Vol. 8:18–44; all in Amgen v. Chugai files.

8. Rasmussen-Davies; Davies testimony, August 17, 1989, 8:23–27.

9. Testimony of Dr. Axel Ullrich, August 16, 1989, Transcripts Vol. 7:12, 17, 67; Deposition transcript (with summaries) from ITC hearings, Eugene Goldwasser, June 17, 1988:59–61, 217–18; both in Amgen v. Chugai files; Hughes-Rathmann, 65; Maniatis T, Fritsch E, and Sambrook J, *Molecular cloning: a laboratory manual* (Cold Spring Harbor, New York: Cold Spring Harbor Laboratory Press, 1982):227–28 et passim.

10. Testimony of Dr. Richard Flavell, September 27, 1989, Transcripts Vol. 27:16–18;

Davies testimony, August 17, 1989, 8:28–34; Goldwasser ITC deposition, June 17, 1988, at 14 and 16; Sue JA and Sytkowski AJ, Site specific antibodies to human erythropoietin directed toward the NH2-termidnal region, *PNAS* 1983;80:3651–55.

11. Stryer L, *Biochemistry* 2nd ed. (San Francisco: WH Freeman, 1981):629; Suggs S, Wallace RB, Hirose T, et al., Use of synthetic oligonucleotides as hybridization probes: isolation of cloned cDNA sequences for human beta-2 microglobulin, *PNAS* 1981;78:6613–17.

12. Maniatis T, Hardison RC, Lacy E, et al., The isolation of structural genes from libraries of eucaryotic DNA, *Cell* 1978;15:667–701; Lawn RM, Fritsch EF, Parker RC, et al., The isolation and characterization of linked delta- and beta-globin genes from a cloned library of human DNA, *Cell* 1978;15:1157–74; Maniatis et al., *Molecular cloning*, chaps. 9–10.

13. Sue and Sytkowski, Site specific antibodies; Flavell testimony, September 27, 1989, 27:48–53; Davies testimony, August 17, 1989, 8:28–34.

14. Alan Levine to Peter Dukes, February 25, 1982, Defendant's Exhibit 910, attached to Defendants' Joint Appendices to Proposed Findings of Fact and Conclusions of Law, Vol. 3:II–102; Amgen v. Chugai files. The 23rd American Society of Hematology meeting in San Antonio took place December 5–8, 1981, and Mach's NIH request probably followed Goldwasser's disclosure of the Hewick sequence there. Sue and Sytkowski, Site specific antibodies, contained a 26 amino acid sequence disclosed at the meeting. Flavell testimony, September 29, 1989;29:62.

15. Rathmann, Biotechnology startups:52–53. Fu-Kuen L and Paddock GV, Characterization of duck genome fragments containing beta and epsilon globin genes, *Gene* 1984;31:59–64. Testimony of FK Lin, August 11, 1989, Transcripts Vol. 4:22–25; Deposition of Carlton Paul, May 18, 1989, Designated Deposition Transcripts and Summaries; Martin Cline notebook extract, October 28, 1981, Plaintiff's Exhibit 170B, attached to Defendants' Joint Appendices to Proposed Findings of Fact and Conclusions of Law, Vol. II–III, II–72; all in Amgen v. Chugai files.

16. Lin to Caruthers, September 24, 1981, Defendants' Exhibit 626; Lin Probe chart, Defendants' Exhibit 750; Rathmann meeting notes of December 1982, Defendants' Exhibit 225, all attached to Defendants' Joint Appendices to Proposed Findings of Fact and Conclusions of Law; Lin testimony, August 11, 1989, Transcripts, Vol. 4:58–59; Lin testimony, August 14, 1989, Vol. 5:102, 110–13; Lin testimony, August 15, 1989, Transcripts Vol. 6:34, 128–29; Lin testimony of August 16, 1989, Transcripts Vol. 7:12, 17, 67; all in Amgen v. Chugai files. On molecular biologists relying on selection for expected events rather than detailed understanding of their experimental systems so as to interpret actual events, see Knorr-Cetina, K, *Epistemic cultures: how the sciences make knowledge* (Cambridge, MA: Harvard University Press, 1999).

17. Anon, Marketing research, *Washington Post*, November 7, 1980:A18; Hilts PJ, Ivy-covered capitalism: universities, in face of faculty fire, weigh founding own firms to cash in on lab work, *Washington Post*, November 10, 1980:A1; Sanger DE, Business rents a lab coat and academia hopes for the best, *NYT*, February 21, 1982:E7; Anon, Coast teachers warned on ties to corporations, *NYT*, February 8, 1983:A16.

18. Hewick interviewed with Maniatis in March 1981 and was hired in April. Hewick testimony, Vol. 11:37–38; Testimony of Edward Fritsch, September 25, 1989, Transcripts Vol. 25:118–28, 134; Testimony of Tom Maniatis, October 18, 1989, Transcripts Vol. 36:57; all in Amgen v. Chugai files. Maniatis et al., Molecular cloning; Rasmussen-Fritsch; Rasmussen-Maniatis.

19. Fritsch probe chart, Defense Exhibit 886, attached to Defendants' Joint Appendices to Proposed Findings of Fact and Conclusions of Law, II–35; Testimony of Edward Fritsch, September 26, 1989, Transcripts Vol. 26:34–35; Hewick lab notebook sequence entry; all in Amgen v. Chugai files; Rasmussen-Fritsch.

20. Lin testimony, Vol. 4:58–59, Vol. 5:111–13, Vol. 6:34; N. Stebbing to E. Goldwasser, December 2, 1982, Defendant's Exhibit 407, attached to Defendants' Joint Appendices to Proposed Findings of Fact and Conclusions of Law, II–91; all in Amgen v. Chugai files.

21. Lin Probe chart; Lin testimony, Vol. 5, 120–21, Vol. 6:24–25; FK Lin, Request form for oligonucleotides, April 26, 1983, Defendants' Exhibit 629–39, attached to Defendants' Joint Appendices to Proposed Findings of Fact and Conclusions of Law, II–92; all in Amgen v. Chugai files. Amino acid sequence from Lai PH, Everett, R, Wang FF, et al., Structural characterization of human erythropoietin, *JBC* 1986;261:3116–19.

22. Rathmann-Hughes:64–65 (quotes); Rasmussen-Vapnek; see also Binder G, *Science lessons: what the business of biotech taught me about management* (Boston: Harvard Business School Publishing), chap. 4; Alegretti D, Epo Timetable, July 27, 1989, Plaintiff's Exhibit 832, with Defendants' Joint Appendices to Proposed Findings of Fact and Conclusions of Law (loose); Lin Probe chart; Lin testimony, Vol. 4:55; all in Amgen v. Chugai files. Re announcement: Fritsch interview; also the reactions of GI and Chugai management at this time confirm that there must have been some such announcement: Yang MC to Sadahiro R, January 16, 1984, Plaintiff's Exhibit 683, Plaintiff's Appendix to Proposed Findings of Fact and Conclusions of Law IX–2, Amgen v. Chugai files. See Lin FK, DNA sequences encoding erythropoietin, US Patent 4,703,008, issued October 27, 1987, on application of November 30, 1984 (continuing application of December 1983).

23. Testimony of Thomas Wall, October 19, 1989, Transcripts, Vol. 37:43–47, 52, 60–62; Erythropoietin distribution data June 1978–present (NIH), December 22, 1983, Plaintiff's Exhibit 344, attached to Defendants' Joint Appendices to Proposed Findings of Fact and Conclusions of Law Appendix with attached documents:Vol 3-II-98; both in Amgen v. Chugai files.

24. Daniel Vapnek, personal communication, November 2102. Vapnek D, Minutes of Epo meeting March 27, 1984, dated April 4, 1984, Affidavits Submitted on Behalf of Defendants Memorandum in Opposition to Plaintiff's Motion for Preliminary Injunction, filed February 11, 1989, Exhibits; Browne J, November 28 MC Epo PDT meeting minutes, dated November 30, 1984, Defendants' Exhibit 349, Defendants' Joint Appendices to Proposed Findings of Fact and Conclusions of Law, Appendix IV–22 (quote); Defendants Summary, ITC Deposition of Jeffrey Browne, May 20, 1988 (admitted as evidence October 18, 1989), Designated Transcripts and Summaries; all in Amgen v. Chugai files. See Lin patent 4,703,008, especially ex. 7, 10. Epo protein made in bacteria and yeast would lack the correct sugars, but the reduced biological activity might be compensated for by a higher dosage of the protein, whereas monkey Epo might work well enough in people and could possibly be converted to the human amino acid sequence later through small, deliberate mutations.

25. Klausner A, "Adjustment" in the blood fractions market, *Bio/Technology* 1985; 3(2):119–25; Davies testimony, Vol. 8:21–44 (quote on 43); Flavell testimony, Vol. 27:96–97 and Vol. 29:60–62; Rasmussen-Davies.

26. Rasmussen-Fritsch; Bishop J, Blood pressure: biotech firms race to market a protein hemophiliacs need, *WSJ*, July 25, 1985:1; Toole J Jr and Fritsch E, Human factor VIII:

C gene and recombinant methods for production, US Patent 4,757,006, issued July 12, 1988, on application of October 28, 1983.

27. Rasmussen-Fritsch; Yang to Sadahiro, January 16, 1984. MC Yang to Takaji Miyake, February 6, 1984, attached to Defendants' Joint Appendices to Proposed Findings of Fact and Conclusions of Law, Appendices, II–58; Testimony of Gabriel Schmergel testimony, October 16, 1989, Transcripts, Vol. 36:35–36 (quoting translation of August 1985 document, Plaintiff's Exhibit 335); both in Amgen v. Chugai files.

28. Testimony of Rodney Hewick, August 23, 1989, Transcripts, Vol. 11: 39–44, and September 19, 1989, Transcripts, Vol. 21: 53–55; Testimony of Edward Fritsch, September 26, 1989, Transcripts, Vol. 26:104–6; all in Amgen v. Chugai files. GI was fortunate in obtaining fetal liver from exactly the developmental stage where Epo transcription peaks, thanks to a Scientific Board member affiliated with Brigham and Women's Hospital. This was a different RNA source from one boasted by GI to Chugai in February, from a liver cancer cell line, likely from Golde; Rasmussen-Fritsch. See Jones S, Hewick R, Fritsch E, et al., Isolation and characterization of genomic and cDNA clones of human erythropoietin, *Nature* 1985;313:806–10; Fritsch E, Jacobs K, and Hewick RM, Method for production of erythropoietin, European Patent 0411678B2, filed December 3, 1985. The last patent covers production of Epo from CHO cells transformed with a human cDNA as described by the *Nature* article, as well as CHO expression vectors, and presumably includes matter covered in GI's unsuccessful US patent applications filed earlier. For basic studies of gene regulation using the commercial Epo clones, see Semenza GL, Dureza RC, Traystman MD, et al., Human erythropoietin gene expression in transgenic mice: multiple transcription initiation sites and cis-acting regulatory elements, *Mol Cell Biol* 1990;10:930–38; Beru N, Smith D, and Goldwasser E, Evidence suggesting negative regulation of the erythropoietin gene by ribonucleoprotein, *J Biol Chem* 1990;265:14100–4.

29. Hewick R and Seehra J, Method for the purification of erythropoietin and erythropoietin compositions, US Patent 4,677,195, issued June 30, 1987, on application of January 11, 1985; Recny MA, Scoble HA, and Kim Y, Structural characterization of natural human urinary and recombinant DNA-derived erythropoietin: identification of des-arginine 166 erythropoietin, *JBC* 1987;262:17156–63.

30. Beck AK, Withy RM, Zabrecki JR, and Masiello MC, Cell encoding recombinant human erythropoietin, US Patent 4,954,437, issued September 4, 1990, on application of September 15, 1986; Powell J, Human erythropoietin gene: high level expression in stably transfected mammalian cells, US Patent 5,688,679, issued November 18 1997, on application of October 6, 1993, continuing abandoned application of June 1986; Powell JS, Berkner KL, Lebo RV, Adamson JW, Human erythropoietin gene: high level expression in stably transfected mammalian cells and chromosome localization, *PNAS* 1986;83:6465–69. First Declaration of Jerry S Powell, unsigned 1993 carbon, Exhibits to the Declarations of Dr. Arthur J Sytkowski, Dr. Jerry S Powell, Lawrence H Thomson, Vol. 1, Exhibit A23, Amgen v. Elanex files. Stipp D and Harris RJ Jr, Amgen gets patent for drug process, sues three rivals, *WSJ* October 29, 1987:11; Anon, Amgen named in countersuit, *NYT*, October 30, 1987:D4; Hughes-Kleid, 148. On the broader topic of both tort and intellectual property lawsuits as contributors to science and industrial policy see Jasanoff S, *Science at the bar: law, science, and technology in America* (Cambridge, MA: Harvard University Press, 1997). Amgen later sued Elanex.

31. Chahine K, Elanex battles for a piece of the erythropoietin market, *NBT* 1998;16:16;

van Brunt J, Epo: biotech's next blockbuster drug? *Bio/Technology* 1987;5:199; Klausner A, J & J's biotech collaborations: a mixed bag, *Bio/Technology* 1988;6:271–75; Revkin A, Firm may have key to treating anemia: hormone process holds promise for kidney ills, *Los Angeles Times*, January 11, 1987:WS10; Stavro B, Promising drug enhances tiny Amgen's reputation, *Los Angeles Times*, May 12, 1987:B1; Pollack A, Gene-splicing payoff is near, *NYT*, June 10, 1987:D1; Schmeck H Jr, Synthesized drug eases kidney ills, *NYT*, January 8, 1987:21.

32. Affidavit of Tuan-Ha Ngoc, January 31, 1989, Affidavits Submitted on Behalf of Defendants Memorandum in Opposition to Plaintiff's Motion for Preliminary Injunction, filed February 11, 1989; Amgen v. Chugai files. Rasmussen-Fritsch; Moskowitz DB, Patent owners gaining clout: challenging patents a riskier business, *Washington Post*, July 15, 1985:WB7.

33. Moxon GW Jr, Products of nature: the new criteria, *Cath U Law Rev* 1971;20:783–90. Jon Harkness has argued that the 1911 Adrenalin decision, so influential on and profitable for the American pharmaceutical industry, reflected a frank misunderstanding of patent law at the time; Harkness J, Dicta on Adrenalin(e): myriad problem with Learned Hand's product-of-nature pronouncements in Parke-Davis v. Mulford, *J Patent Trademark Office Soc* 2011;93:363–99; Anon, Genentech, Inc. and Wellcome plc sue each other over right to market tPA, *BLR* 1987;6:200–201; Anon, Genentech loses patent dispute over tPA in Britain's High Court, *BLR* 1987;6:282–339; Hamilton J. Key precedent for biotech patents may fall in wake of Amgen-GI Decision, *Bioworld*, December 14, 1989 (unpaginated newswire).

34. Van Brunt J, Amgen marches to the clinic, *NBT* 1987;5:892–98; Bates J, Biotech patent gives Amgen an edge, but legal battles lie ahead, *Los Angeles Times*, November 3, 1987:B9A; Wessel D, Genetics Institute gets patent on drug that fights anemia in kidney patients, *WSJ*, July 2, 1987:22; Jack Bowman to George Rathmann, February 26, 1988; Defendants' Exhibit 818, attached to Defendants' Joint Appendices to Proposed Findings of Fact and Conclusions of Law, Appendix, Vol. 3, XVII–6, Amgen v. Chugai files.

35. Rundle RL, Amgen, cleared to sell kidney-patient drug, still faces big hurdles, *WSJ*, June 2, 1989:A1; idem, Row with co-marketer adds to Amgen's woes, *WSJ*, February 6, 1989:B6; Bernstein JA, District court holds that Amgen infringes genetics institute erythropoietin patent, *BLR* 1989;8:78–81.

36. Ullrich testimony, August 16, 1989, Vol. 7, 15; Magistrate Patti Saris, Memorandum and Order, Amgen Inc. v. Chugai Pharmaceutical Co. Ltd. and Genetics Institute, Inc. C.A. No. 87-2617-Y, United States District Court, Massachusetts District, December 11, 1989, reproduced in *BLR* 1990;9:25–71 ("close").

37. Saris, Memorandum and Order; Ducor P, *Patenting the recombinant products of biotechnology and other molecules* (London: Kluwer, 1998):47–48. Regarding DNA diagnostics, I refer to the problem captured by the Myriad Genetics patents on BRCA genes, which largely hinged on the patentability and definition of products of nature; Harkness, Dicta; Cook-Deegan R, Law and science collide over human gene patents, *Science* 2012;338:745–47; Kevles D, Genes, railroads and regulations: intellectual property and the public interest, in Biagioli M and Riskin J, eds., *Nature engaged: science in practice from the Renaissance to the present* (London: Palgrave Macmillan, 2012):147–62; Liptak, A, Justices, 9–0, bar patenting human genes, *NYT*, June 14, 2013:701.

38. Andrews EL, Ruling may hurt Amgen's rights to drug, *NYT*, December 18, 1989: D1.

39. Stipp D, Genetics Institute, Japanese firm seek injunction against Amgen in patent case, *WSJ*, January 31, 1990:B2; Anon, Amgen wins ruling in its patent dispute, *WSJ*, April 18, 1990:A12 ("delay tactic"); Genetics Institute is set back in bid for approval of drug, *WSJ*, January 15, 1991:B7; Leary WL, New anemia drug approved by us for kidney disease: genetically engineered, *NYT*, June 2, 1989:A1; Anon, Amgen's Epogen Medicare/Medicaid reimbursements will approach $200 mil. in 1990, chairman Rathmann estimates at Rep. Waxman's Feb. 7 orphan drug hearing, *FDC Reports*, February 12, 1990:5; Griffiths RI, Powe NR, Greer J, et al., *Health Care Financing Rev*1994;15:83–102.

40. Asbury CH, The Orphan Drug Act: the first 7 years. *JAMA* 1991;265:893–97; Shulman SR, Bienz-Tadmor B, Seo PS, et al., Implementation of the Orphan Drug Act: 1983–1991, *Food & Drug Law J* 1992;47:363–403; Gladwell M, Waxman ends effort to tighten provisions of Orphan Drug Act, *Washington Post*, March 9, 1988:F1; Yoder SK, Trade association for biotechnology becomes unspliced, *WSJ*, March 28, 1990:B8; Gladwell M, Drug law splits biotech firms: critics in industry, Congress question winner-take-all effect, *Washington Post*, April 1, 1990:H1 ("true orphans"); Gladwell M, Bill seeks more drug competition: authors want to weaken monopolies for rare-disease treatments, *Washington Post*, May 13, 1990; H4 ("covenant"); Editorial, Waxman threatens orphans, *WSJ*, May 8, 1990:A24; Anderson J and van Atta D, "Orphan Drug" law: license to profiteer?, *Washington Post*, August 2, 1990:B13; Anon, Bush pocket-vetoes bill to spur orphan drug development, *Washington Post*, November 10, 1990:A13.

41. Decision of Judge Lourie, United States Court of Appeals, Federal Circuit, March 5, 1991, in Amgen, Inc., Plaintiff/Cross-Appellant, v. Chugai Pharmaceutical Co., Ltd., and Genetics Institute, Inc., Defendants-Appellants, Nos. 90-1273, 90-1275, 927 F.2d 1200, 59 USLW 2575, 18 U.S.P.Q.2d 1016, at 41–51; Warburg RJ, Recombinant conception: does it require the DNA? *Bio/Technology* 1993;11:1306–7. On the consistency of Lourie's decisions with chemical precedent, see Storella J, *Amgen, Fiers, and Bell:* the Federal Circuit and the patentability of genes, *BLR* 1994;13:459–67.

42. Greenfield MS, Recombinant DNA technology: a science struggling with the patent law, *Stanford Law Rev* 1992;44:1051–94; this author, however, seems to presuppose that innovation in cloning is more valuable to the public than innovation in protein chemistry, by proposing only remedies that would make it easier to protect the genetic side of biotechnology drug research and development; decision of Judge Lourie at 107–15.

43. Hewick contended that the measures should be similar because his procedure preserved the sugar groups that make up fully half the natural hormone's weight, but are removed by other purification methods. On Epo sales and profits at the time of the CAFC decision, see: Anon, Amgen Inc. reports profit of $60 million for fourth quarter, *WSJ*, February 4, 1992:C18; Anon, Amgen gives backing to forecasts of rise in quarterly earnings, *WSJ*, September 22, 1992:C1. Linsenmeyer A, No Band-Aid solution, *Financial World*, January 21, 1992:25. On the perception and impact of the decision, see: Rundle R and Stipp S, Amgen wins biotech drug patent battle, *WSJ*, March 7, 1991:A3–A5; Ratner M, EPO, Factor VIII shockers, *Bio/Technology* 1991;9:327.

44. I am convinced by appellants' argument that in rejecting the Hewick patent on enablement grounds Judge Lourie confused the specific activities of Miyake Shipment 3 before Hewick's application of RP-HPLC with its activity after. Also, his decision ignored both the success of expert witness Kung in obtaining Epo with in vitro specific activity

greater than 170,000, and Goldwasser's testimony that RP-HPLC-purified Epo with activity greater than 80,000 showed, in his own hands, virtually identical in vitro and in vivo activity measures. Defendant-Appellants' Joint Petition for Rehearing and Joint Suggestion for Rehearing *in Banc*, CAFC Appeal Nos. 90-1273-75, 19 March 1991, 4–8 et passim (document gift of Bruce Eisen).

45. Dutfield G, *Intellectual property rights and the life science industries: past, present and future*, 2nd. ed. (Singapore: World Scientific, 2009):142 et passim; Rowland B and Field B, *In re Wright:* a new standard of enablement for biotechnology, *BLR* 1995;14:295–311 ("fetish" on 309, "elect to invest" on 305). I am indebted to Steve Hubicki for insight about the obviousness implications of the Amgen appeals decision.

46. Bent S and Sandercock C, CAFC to PTO: "it's clear as a *Bell*." *BLR* 1995;14:351–53 ("salami" on 353); Mintz M, Still hard to swallow, *Washington Post*, November 2, 2001:B1; Lasagna L, Congress, the FDA, and new drug development: before and after 1962, *Perspect Biol Med* 1989;32:322–43; Hilts P, *Protecting America's health: the FDA, business, and one hundred years of regulation* (New York: Knopf, 2003), chap. 9.

47. Office of Technology Assessment, *Commercial biotechnology: an international analysis*, OTA-BA-218 Washington, DC: US Congress, 1984:150. Anon, A groundswell of drugs from the biotech pipeline, *Chemical Week*, January 21, 1987;6. On interpretive capture see Dutfield, *Intellectual property rights*, chap. 3.

48. A similar observation has been made about the CAFC's implicit policy across numerous high technology sectors; Jaffey AB and Lerner J, *Innovation and its discontents: how our broken patent system is endangering innovation and progress, and what to do about it* (Princeton: Princeton University Press, 2011). On the OTA's closure, see Cooper K, Lawmakers ready to shut technology office, *Washington Post*, June 12, 1995:A17; Granger M, Death by Congressional ignorance, *Pittsburgh Post-Gazette*, August 2, 1995:A1; Langreth R, Dark days for science?, *Popular Science*, October 1995, 74–80. The following reports all focus on how to extract national benefit from genetic engineering; all are authored by the Congressional Office of Technology Assessment, published by the Government Printing Office, Washington, DC, and available for download at http://www.fas .org/ota/otareports/topic/btopics: *Impacts of applied genetics: micro-organisms, plants, and animals* (April 1981); *Commercial biotechnology: an international analysis* (January 1984); *Human gene therapy* (December 1984); *New developments in biotechnology: ownership of human tissues and cells* (March 1987); *New developments in biotechnology: public perceptions of biotechnology* (May 1987); *Mapping our genes-genome projects: how big? how fast?* (April 1988); *New developments in biotechnology: field-testing engineered organisms: genetic and ecological issues* (May 1988); *New developments in biotechnology: U.S. investment in biotechnology* (July 1988); *New developments in biotechnology: patenting life-special report* (April 1989); *Biotechnology in a global economy* (October 1991); *Federal technology transfer and the human genome project* (September 1995).

49. The European Patent Office rejected the European version of the Hewick pure Epo patent on a very narrow procedural technicality, after it had survived vigorous opposition by Amgen. See Eisen B, Sudden death in Munich: a pitfall in filing via the EPO, *BLR* 1998;7:733–36. On the European treatment of the Epo DNA patents, see Maebius SB and Sandercock CG, August brings good news and bad news for Amgen, BLR 1995;14:609–23; Chahine, Elanex battles; Dove A, Betting on biogenerics, *NBT* 2001;19:117–20; Firn D,

Baxter buys up anaemia drug, *Financial Times* (London), October 17, 2001:32. Fritsch, Jacobs, and Hewick, Method for the production of erythropoietin (European patent 0411678B2).

50. Newman L, Avalanche of direct-to-consumer drug marketing brings new questions, *J Natl Cancer Inst* 2000;92:964–67; Napoli M, The FDA is taking a hard look at anemia drugs for people with cancer, *HealthFacts*, June 2007:5. For the politically handicapped efforts of FDA staffers to stop Johnson & Johnson's off label fatigue and high dose marketing, see Sharp K, *Blood feud: the man who blew the whistle on one of the deadliest prescription drugs ever* (New York: Dutton, 2011):223–27, 253–56.

51. Melnikova I, Anaemia therapies, *Nature Rev Drug Discovery*2006;5:627–28; Greb E and van Arnum P, Pharmaceutical Technology's manufacturers' rankings, *Pharmaceutical Technology*, June 2007, 54–69. Steinbrook R, Erythropoietin, the FDA, and oncology, *NEJM* 2007;356:2448–51; Anon, How have safety concerns and new warnings affected the anemia market?, M2 Presswire, August 19, 2008 (unpaginated news wire); Repetto L, Moeremans K, Annemans L; European Organisation for Research and Treatment of Cancer, European guidelines for the management of chemotherapy-induced anaemia and health economic aspects of treatment, *Cancer Treat Rev* 2006;32(Suppl 2):S5–9; Bennett CL, McKoy JM, Henke M, et al., Reassessments of ESAs for cancer treatment in the US and Europe, Oncology (Williston Park) 2010;24:260–68.

52. Editorial, Epoeitin: for better or worse?, *Lancet Oncol*2004;5:1. Henke M, Laszig R, Rübe C, et al., Erythropoietin to treat head and neck cancer patients with anaemia undergoing radiotherapy: randomised, double-blind, placebo-controlled trial, *Lancet* 2003;362: 1255–60; Leyland-Jones B, Semiglazov V, Pawlicki M, et al., Maintaining normal hemoglobin levels with epoetin alfa in mainly nonanemic patients with metastatic breast cancer receiving first-line chemotherapy: a survival study, *J Clin Oncol* 2005;23:5960–72; Bennett CM, Silver SM, Djulbegovic B, et al., Venous thromboembolism and mortality associated with recombinant erythropoietin and darbepoetin administration for the treatment of cancer-associated anemia, *JAMA* 2008;299:914–24; Bohlius J, Schmidlin K, Brillant C, et al., Recombinant human erythropoiesis-stimulating agents and mortality in patients with cancer: a meta-analysis of randomised trials, *Lancet* 2009;373:1532–42; European Medicines Agency, EMEA recommends a new warning for epoetins for their use in cancer patients, Press release, June 26, 2008, Doc. EMEA/CHMP/333963/2008 -corr. (http://www.emea.europa.eu/docs/en_GB/document_library/Press_release/2009/ 11/WC500015069.pdf); Bennett, McKoy, and Henke, Reassessments of ESAs for cancer. In the United States, ESA usage increased 340% between 2001 and 2006, and decreased 60% since 2007. In Europe between 2001 and 2006, ESA use increased 51%; since 2006, use decreased by 10%.

53. Bennett CL, Lai SY, Henke M, et al., Association between pharmaceutical support and basic science research on erythropoiesis-stimulating agents, *Arch Internal Med* 2010;170;1490–98. The mechanisms behind the enhanced likelihood of company-sponsored clinical trials to report favorable outcomes, such as sponsorship of more trials than will be published, also cost a great deal of money and therefore might also be attributed to excessive profit margins.

Chapter 6 · tPA

1. Schroeder J, Current status of cardiac transplantation, 1978, *JAMA* 1979;241(19): 2069–71; Illich I, *Limits to medicine: medical nemesis—the expropriation of health* (London: Marion Boyars, 1976): 105–6. For a classic study capturing the controversial status of artificial organs at the time, see Fox RC and Swazey JP, *The courage to fail—a social view of organ transplants and dialysis* (Chicago: University of Chicago Press, 1974). The critique of artificial hearts appears to have led to a review of the NIH program: Reeves TJ, Blinks JR, Cox JR, et al., Mechanically assisted circulation—the status of the NHLBI program and recommendations for the future, *Artificial Organs* 1977;1:39–58; McIntosh HD and Garcia JA, The first decade of aortocoronary bypass grafting, 1967–1977: a review, *Circulation* 1978;57:405–31; Gruntzig AR, Senning A, Siegenthaler WE, Nonoperative dilation of coronary artery stenosis: percutaneous transluminal coronary angioplasty, *NEJM* 1979;301:61–68. On epidemiological trends in heart disease see, National Heart Lung and Blood Institute, *2009 NHLBI morbidity and mortality chart book* (Washington, DC: NIH, 2009 (http://www.nhlbi.nih.gov/resources/docs/cht-book.htm).

2. Gonzalez E, Intracoronary thrombolysis to abort heart attacks: wave of the future?, *JAMA* 1981;245:11–13; Laffel GL and Braunwald E, Thrombolytic therapy: a new strategy for the treatment of acute myocardial infarction (1), *NEJM* 1984;311: 710–17; Laffel GL and Braunwald E, Thrombolytic therapy: a new strategy for the treatment of acute myocardial infarction (2), *NEJM* 1984;311:770–76.

3. Marks HM, *The progress of experiment: science and therapeutic reform in the United States, 1900–1990* (Cambridge: Cambridge University Press, 1997); Abraham J, *Science, politics and the pharmaceutical industry* (London: University College Press, 1995); Carpenter C, *Reputation and power: organizational image and pharmaceutical regulation at the FDA* (Princeton: Princeton University Press, 2010). Here I am suggesting with the *pas de trois* concept, specifically in the matter of trial design (and not the other predictors of sponsorship bias in the clinical trial literature), that it is not as easy to separate expert (or "key opinion leader") medical "interest" from sponsor commercial "interest" in the sponsored trials in which both actively participate as many well-meaning critiques suppose. For a recent review also noting the difficulty, see Sismondo S, Corporate disguises in medical science: dodging the interest repertoire, *Bull Sci Tech Society* 2011;31:482–92; for an exploration of the history of convergent interests in trial design, see Rasmussen N, The drug industry and clinical research in interwar America: three types of physician collaborator, *Bull Hist Med* 2005;79:50–80.

4. Anon, Abbott first to be granted US marketing approval for urokinase, *FDC Reports,* January 23, 1978:TG1; Klausner A, Activating the body's blood clot dissolvers: biotech's new role, *Bio/Technology,* June 1983:330–36.

5. Astrup T and Permin PM, Fibrinolysis in the animal organism, *Nature* 1947;159:681–82; Astrup T, Tissue activators of plasminogen, *Fed Proc* 1966;25:42–51; Kwaan HC, The life and vision of a pioneer: a tribute to Tage Astrup, *Semin Thromb Hemost* 2007;33:299–300; Sherry S, Fibrinolysis, *Ann Rev Medicine* 1968;19:247–65; Collen D, On the regulation and control of fibrinolysis: Edward Kowalski Memorial Lecture, *Thromb Haemost* 1980;43:77–89.

6. While Goeddel recalls that Genentech discussed tPA and urokinase as cloning targets before the Gruenenthal contract, prior interest in either protein does not fit Heyneker's recollection; Hughes-Goeddel, 56, 70; Hughes-Heyneker, 85, 124–27.

7. Hughes-Pennica, 12–13; Lynch KR, Pennica D, Ennis HL, et al., Separation and purification of the mRNAs for vesicular stomatitis virus NS and M proteins, *Virology* 1979;98:251–54; idem, Temporal regulation of the rate of vesicular stomatitis virus mRNA translation during infection of Chinese hamster ovary cells, *Virology* 1981;108: 277–85; Collen D, Human tissue-type plasminogen activator: from the laboratory to the bedside, *Circulation* 1985;72:18–20; Rijken DC and Collen D, Purification and character-ization of the plasminogen activator secreted by human melanoma cells in culture, *JBC* 1981;156:7035–41; Rifkin DB, Loeb JN, Moore G, et al., Properties of plasminogen acti-vators formed by neoplastic human cell cultures, *J Exp Med* 1974;139:1317–28; Hughes-Kleid, 129–30; Hughes-Pennica, 13–20.

8. Pennica D, Holmes WE, Kohr WJ, et al., Cloning and expression of human tissue-type plasminogen activator cDNA in *E. coli.*, *Nature* 1983;301:214–21; Jacobs K, Shoe-maker C, Rudersdorf R, et al., Isolation and characterization of genomic and cDNA clones of human erythropoietin, *Nature* 1985;313:806–10.

9. Goeddel D, Kohr WJ, Pennica D, Vehar GA, Human tissue plasminogen activa-tor, US Patent 4,766,075, issued August 23, 1988, on application of April 7, 1983, and idem, Human tissue plasminogen activator, US Patent 4,853,330, issued August 1, 1989, on application of April 21, 1988, both continuing an application first filed May 5, 1982. Pennica, Holmes, and Kohr, Cloning and expression; Hughes-Goeddel, 71–72; Hughes-Pennica, 17–20. Pennica evidently recalled in an interview with journalist Linda Marsa that she first saw sequencing evidence of a clone on October 25, 1981; Marsa L, *Prescrip-tion for profits* (New York: Scribner, 1997):130. Goeddel declaration of December 21, 1987, re patent application 06/483,052, Defendant's Exhibit 816, in Genentech v. Wellcome files.

10. Pennica, Holmes, and Kohr, Cloning and expression; Hughes-Kleid, 134–35; Hughes-Goeddel, 71; Goeddel declaration; Genentech, TPA produced by recombinant DNA techniques (press release), July 28, 1982 (http://www.gene.com/gene/news/press-releases/display.do?method=detail&id=4185); testimony of Tom Maniatis, March 26, 1990, Transcript of Proceedings:1851–52, Genentech v. Wellcome files.

11. Hughes-Pennica, 20–23; Pennica, Holmes, and Kohr, Cloning and expression; Edlund T, Ny T, Ranby M, et al., Isolation of cDNA sequences coding for a part of human tissue plasminogen activator, *PNAS* 1983;80:349–52. If editorial conspiracy explains the simultaneous publication, it follows a tradition at least as old as Darwin's and Wallace's joint publication in *J Proc Linn Soc* 1858;3:45–62, and further attests to Genentech's hon-orary academic standing.

12. Hughes-Pennica, 21; Goeddel declaration; Goeddel, Kohr, Pennica, and Vehar, US patents 4,766,075 and 4,853,330.

13. Banville D, Kay RM, Harris R, et al., The nucleotide sequence of the mRNA encod-ing a tadpole beta-globin polypeptide of *Xenopus laevis*, *JBC* 1983;258:7924–27. First Affi-davit of Mary-Jane Gething, April 14, 1987, Plaintiff's Exhibit 433 and First Affidavit of Joseph Frank Sambrook, April 14, 1987, Plaintiff's Exhibit 386, both in Genentech v. Wellcome records. Rasmussen-Kay; Rasmussen-Larsen.

14. Kaufman RJ, Wasley LC, Spiliotes AJ, et al., Coamplification and coexpression of human tissue-type plasminogen activator and murine dihydrofolate reductase sequences in Chinese hamster ovary cells, *Mol Cell Biol* 1985;5:1750–59. For cloning details also see Kaufman RJ, Method for making tissue plasminogen activator, US Patent 5,079,159,

issued January 7, 1992, on application of April 25, 1988 (continuing an application of December 27, 1983), and idem, Vectors and methods for transformation of eucaryotic cells, US Patent 4,740,461, issued April 26, 1988, on application of December 27, 1983. The construction of complete cDNA p205 is described in the latter, and p205's use in confirming tPA protein translation (possibly a fusion protein) in COS cells in mid-January 1983 is described in Gething affidavit.

15. On final confirmation of DNA sequence, Lawrence Flynn, memo to file, May 18, 1983, Plaintiff's Exhibit 625, Genentech v. Wellcome files. The delay between detection of a clone expressing tPA protein in January and confirmatory sequencing in May 1983 may reflect mutations subsequently introduced in the first isolate, requiring reassembly and recloning. Rasmussen-Kay; Rasmussen-Fritsch interview. The British tPA decision and the Saris decision in Amgen v. Chugai upset common expectations as to patent scope regarding first-generation products; see Hamilton J, Key precedent for biotech patents may fall in wake of Amgen-GI decision, *Bioworld*, December 14, 1989 (unpaginated newswire). Compare Kaufman's vector p7B1 in US patent 5,079,159 with the pETPER vector described in Goeddel, Kohr, Pennica, and Vehar, US Patent 4,853,330.

16. Anon, Wellcome/Genetics biotech company, *Scrip*, September 17, 1986:6; Anon, Genetics Institute/Burroughs Wellcome tPA is in phase I in 30 centers: biotech firm has second generation longer acting tPA in development, *FDC Reports*, October 20, 1986:12; Anon, Genentech will appeal tPA ruling; GI unveils 2nd-generation analog, *Biotechnology Newswatch*, July 20 1987:1. Plaintiff's List of Exhibits, Genentech v. Wellcome, items PX 5, PX 10, PX 477. Rasmussen-Larsen; Hughes-Kleid, 148.

17. US Congress, Office of Technology Assessment, *Commercial biotechnology: an international analysis* (OTA-BA-218) (Washington, DC: GPO, 1984):104, 135; Anon, Biogen/SKF tPA agreement, *Scrip*, August 13, 1984:8; Rosa JJ and Rosa MD, Modified tissue plasminogen activators. US Patent 4,753,879, issued June 28, 1988, on application of August 27, 1984; Rasmussen-Rosa; Kresge N, Simoni R, and Hill RL, The development of site-directed mutagenesis by Michael Smith, *JBC* 2006;281:e3.

18. Zamarron C, Lijnen HR, and Collen D, Kinetics of the activation of plasminogen by natural and recombinant tissue-type plasminogen activator, *JBC* 1984;259:2080–83; Van de Werf F, Bergmann SR, Fox KA, et al., Coronary thrombolysis with intravenously administered human tissue-type plasminogen activator produced by recombinant DNA technology, *Circulation* 1984;69:605–10; Collen D, Topol EJ, et al., Coronary thrombolysis with recombinant human tissue-type plasminogen activator: a prospective, randomized, placebo-controlled trial, *Circulation* 1984;70:1012–17; Goldsmith M, Recombinant plasminogen agent continues to show promise in trials, *JAMA* 1985;253:1693–94.

19. Gilbert W, Why genes in pieces?, *Nature* 1978;271:501; Larsen G, Henson K, and Blue Y, Variants of human tissue-type plasminogen activator fibrin binding, fibrinolytic, and fibrinogenolytic characterization of genetic variants lacking the fibronectin finger-like and/or the epidermal growth factor domains, *JBC* 1988;263:1023–29; Collen D, Stassen JM, and Larsen G, Pharmacokinetics and thrombolytic properties of deletion mutants in human tissue-type plasminogen activator in rabbits, *Blood* 1988;71:216–19; Larsen GR, Metzger M, Henson K, et al., Pharmacokinetic and distribution analysis of variant forms of tissue-type plasminogen activator with prolonged clearance in rats, *Blood* 1989;73:1842–50; Hansen L, Blue Y, Barone KM, et al., Functional effects of asparagine-linked oligosaccharide on natural and variant human tissue-type plasminogen activator,

JBC 1988;263:15713–19; Aher TJ, Morris GE, Barone KM, et al., Site-directed mutagenesis in human tissue-plasminogen activator: distinguishing sites in the amino-terminal region required for full fibrinolytic activity and rapid clearance from the circulation, *JBC* 1990;265:5540–45; Rasmussen-Larsen.

20. Collen, Stassen, and Larsen, Pharmacokinetics and thrombolytic properties; Larsen, Metzger, and Henson, Pharmacokinetic and distribution analysis; Hansen, Blue, and Barone, Functional effects; Anon, Suntory introduces the 2nd generation tPA from GI, *Pharma,* December 12, 1988:10; Rasmussen-Larsen.

21. Quotes from Justice Whitford in Anon, Genentech loses patent dispute over tPA in Britain's High Court, *BLR* 1987;6:282–339, except Hughes-Kiley, 68 ("greedy").

22. Sun M, Companies vie over new heart drug, *Science* 1987;237:120–22; Anon, Competing tPA developers worldwide nip at Genentech's heels, *Biotechnology Newswatch,* December 7, 1987:5.

23. Anon, Plasma-purified-protein patent also covers cloned product, US Court rules, *Biotechnology Newswatch,* August 3, 1987:1 ("troublesome"); Anon, Genentech's tPA sinks again in U.K: investors sue firm's officers, *Biotechnology Newswatch,* November 7, 1988:3; Anon, Genentech receives broad tPA patent in the United States, files patent infringement suit against Burroughs Wellcome, *BLR* 1988;7:274–75; Anon, Genentech receives second tPA United States patent—files patent infringement suit against Wellcome, *BLR* 1988;7:357–58; Klausner A, Second generation t-PA race heats up, *NBT* 1987;5:869–70; Anon, SmithKline terminates tPA agreements, *FDC Reports,* July 11, 1988, Briefs (unpaginated); Anon, Wellcome drops Duteplase & NPA, *Scrip,* May 16, 1990:29; Lublin J, Wellcome halts six-year effort on heart drug, *WSJ,* May 11, 1990:B1.

24. Davis M, Follow up: Federal Circuit rules on application of doctrine of equivalents for genetically engineered compositions, *BLR* 1994;13:648–755; Genentech Inc. v. Wellcome Foundation Limited, United States Court of Appeals, Federal Circuit, June 27, 1994, 29 F.3d 1555, 31 U.S.P.Q.2d 1161, Opinion by Plager (with concurrence by Lourie).

25. Genentech, Genentech responds to ruling regarding t-PA patent infringement by Genetics Institute (press release), June 28, 1994 (http://www.gene.com/gene/news/press-releases/display.do?method=detail&id=4413); Anon, Bristol-Myers deal on clot-dissolving drug, *NYT,* October 19, 1995:D4; Anon, "Clot-Buster" drug phase II trials to begin, *Gene Therapy Weekly,* October 30, 1995:7. InTIME-II Investigators, Intravenous nPA for the treatment of infarcting myocardium early; InTIME-II, a double-blind comparison of single-bolus lanoteplase vs accelerated alteplase for the treatment of patients with acute myocardial infarction, *Eur Heart J* 2000;21:2005–13; Rasmussen-Larsen.

26. On the evolution of prescription monitoring and medical marketing and research in the United States, see Greene JA, Pharmaceutical marketing research and the prescribing physician, *Ann Intern Med* 2007;146:742–48, and idem, *Prescribing by numbers: drugs and the definition of disease* (Baltimore: Johns Hopkins University Press, 2007).

27. Marsa, *Prescription for Profits,* chap. 6.

28. TIMI Study Group, The thrombolysis in myocardial infarction (TIMI) trial, phase I findings, *NEJM* 1985;312:932–36; Williams DO, Borer J, Braunwald E, et al., Intravenous recombinant tissue-type plasminogen activator in patients with acute myocardial infarction: a report from the NHLBI thrombolysis in myocardial infarction trial, *Circulation* 1986;73:338–46; Marsa, *Prescription for Profits,* chap. 6.

29. Marsa, *Prescription for Profits,* chap. 6 (quote on 160); Topol EJ, Califf RM, George

BS, et al., Insights derived from the thrombolysis and angioplasty in myocardial infarction (TAMI) trials, *J Am Coll Cardiol* 1988;12(Supp):24A–31A.

30. Anon, Genentech submits to FDA drug to treat heart attack, *WSJ*, April 28, 1986:1; Anon, Genentech is planning to at least double 30-man sales force to 60–80 reps in 1987, *FDC Reports*, October 20, 1986:11; Specter M, Clot-dissolving drug waylaid: tests called inadequate to allow sales, *Washington Post*, May 30, 1987:A6; Editorial, Report from the Commissioners, *WSJ*, June 25, 1987:1. Chase M and Davidson J, FDA to clear Genentech drug for blood clots, *WSJ*, November 13, 1987:1; National Heart Foundation of Australia Coronary Thrombolysis Group, Coronary thrombolysis and myocardial salvage by tissue plasminogen activator given up to 4 hours after onset of myocardial infarction, *Lancet* 1988;8579: 203–8.

31. Pollack A, Fast start for bioengineered drug: marketing strategy Helps T.P.A. sales, *NYT*, January 5, 1988: D1; Anon, Genentech Activase sales approach $1.5 mil. per day in first 45 days of marketing; Protropin sales double to $86 mil in 1987, CEO Swanson says, *FDC Reports*, January 18, 1988:4; Topol EJ, Bates ER, Walton JA Jr, et al., Community hospital administration of intravenous tissue plasminogen activator in acute myocardial infarction: improved timing, thrombolytic efficacy and ventricular function, *J Am Coll Cardiol* 1987;10:1173–77; Genentech, Thrombolytics U.S. sales (http://www.gene.com/gene/about/ir/historical/product-sales/thrombolytic.html).

32. Hughes-Goeddel, 55–56; Chase M, Battle of heart-attack drugs heats up as Smith Kline decries Genentech tactics, *WSJ*, March 8, 1990:B1, B6 ("spirit of competition"); Hill C, Smith Kline faces Genentech suit over heart drugs, *WSJ*, February 26, 1991:A5; Waldholz M, Heart-drug studies to fuel firms' rivalry, *WSJ*, March 1, 1991:B2; King, RT Jr, Profit prescription: in marketing of drugs, Genentech tests limits of what is acceptable, *WSJ*, January 10, 1995:A1.

33. Chase M, Wider uses seen for heart drug, but side effects pose a challenge, *WSJ*, March 23, 1987:1; The International Study Group, In-hospital mortality and clinical course of 20,891 patients with suspected acute myocardial infarction randomised between alteplase and streptokinase with or without heparin, *Lancet* 1990;336:71–75; International Study of Infarct Survival Collaborative Group, ISIS-3: a randomised comparison of streptokinase vs tissue plasminogen activator vs anistreplase, *Lancet* 1992;339(8796):753–70; GUSTO Investigators, An international randomized trial comparing four thrombolytic strategies for acute myocardial infarction, *NEJM* 1993;329:673–82; Ridker PM, O'Donnell C, Marder VJ, and Hennekens CH, Large-scale trials of thrombolytic therapy for acute myocardial infarction: GISSI-2, ISIS-3, and GUSTO-1, *Ann Int Med* 1993;119:530–32; Collins R, Peto R, Baigent C, et al., Aspirin, heparin, and fibrinolytic therapy in suspected acute myocardial infarction, *NEJM* 1997;336:847–60.

34. Grines CL, Browne KF, Marco J, Rothbaum D, et al., A comparison of immediate angioplasty with thrombolytic therapy for acute myocardial infarction, the Primary Angioplasty in Myocardial Infarction study group, *NEJM* 1993;328:673–79; Cucherat M, Bonnefoy E, Tremeau G, Primary angioplasty versus intravenous thrombolysis for acute myocardial infarction, *Cochrane Database Syst Rev* 2003;CD001560; Nabel EG, Braunwald E, A tale of coronary artery disease and myocardial infarction, *NEJM* 2012;366:54–63.

35. Pollack A, U.S. biotechnology leader to sell Swiss 60% stake, *NYT*, February 3, 1990:1; Rigdon JE, Fatal blunder: Genentech CEO, a man used to pushing limit, exceeds it and is out, *WSJ*, July 11, 1995:A1; Abate T, The birth of biotech: how the germ of an

idea became the genius of Genentech, *San Francisco Chronicle*, April 1, 2001:B1; Drews J, Strategic choices facing the pharmaceutical industry: a case for innovation, *Drug Development Today* 1997;2:72–78 (quote p. 76); Hughes-Yansura, 90 ("dark ages"); Hughes-Kleid, 126–27, 146–48 ("the worst thing you ever want to do as a PhD is go work for an MD," 126).

36. Sugawara S, Genex acquired by N.J. company in $7 million stock swap, *Washington Post*, June 13, 1991:B12; Thayer A, Genetics Institute: firm sells 60% share for $666 million, *Chemical & Engineering News*, September 30, 1991:6; Rundle R, Large-scale shakeout looms for biotechnology firms, *WSJ*, April 6, 1994:B4.

37. Wessel D, Biogen N.V. taps James L. Vincent as chief executive, *WSJ*, October 1, 1985: 1; Anon, Biogen NV: Glaxo Holdings to purchase firm's Geneva laboratory, *WSJ*, July 29, 1987:2; Joseph Rosa, unpublished interview with History Company, circa 1989 (courtesy J. Rosa); Chase M, Medicine: Chiron to seek FDA approval of MS drug, *WSJ*, March 4, 1993: B1; Fisher L, A new model for biotechnology, *NYT*, April 4, 1993: A5; Dutton G, Biotech: risky business?, *Management Review* 1995;84:36; Johannes L, New drug aims to win over MS sufferers, *WSJ*, May 20, 1996:B1.

38. Powell WW and Sandholtz K, Chance, nécessité, et naïveté: ingredients to create a new organizational form. In: Padgett JF and Powell WW, eds., *The emergence of organizations and markets* (Princeton: Princeton University Press, 2012):379–433; Hughes-Boyer, 96.

Conclusion · Science, Business, and Medicine in the First Age of Biotech

1. Rasmussen-Fritsch; Rasmussen-Rosa; Rasmussen-Seeburg; Rasmussen-Larsen.

2. For example, in 1987 Joseph Rosa was head of the "protein chemistry group" and "director of protein chemistry" at Biogen; Robert Kay was a molecular genetics lab chief under whom Larsen worked, at first in effect as a postdoc, at Genetics Institute in 1982; Genentech around 1980 had three departments, the "cloners" of the Department of Molecular Biology being more important than the "DNA chemists" or the "sequencers." Joseph Rosa, unpublished interview with History Company, circa 1989 (courtesy Joseph Rosa); Rasmussen-Larsen; Rasmussen-Kay; Hughes-Goeddel, 55; Hughes-Rutter, 74.

3. Chase M, Search for "superbugs": industry sees a host of new products emerging from its growing research, *WSJ*, May 10, 1979:48 (Farley quote). Rasmussen-Rosa ("low hanging fruit"). Examples of young biologists discussed here, whose 1980s work in biotech firms led to academic careers, would include Peter Seeburg and Axel Ullrich of Genentech, and Randall Kaufman of Genetics Institute.

4. Kenney M, *Biotechnology: the university-industrial complex* (New Haven: Yale University Press, 1986); Krimsky S, *Biotechnics and society: the rise of industrial genetics* (New York: Praeger, 1991); Wright S, *Molecular politics: developing American and British regulatory policy for genetic engineering* (Chicago: University of Chicago Press, 1994). That entrepreneurial biologists successfully remade the ethos at their academic institutions does not imply they did so quickly or easily; see Jones M, Entrepreneurial science: the rules of the game, *SSS* 2009;39:821–51.

5. Lane E, Roehr B, and Lempinen EW, As U.S. nears a "fiscal cliff," concerns rise for future R&D, *Science* 2012;336:997. Apart from funding curiosity-driven research more generously I am suggesting that there may be ways of encouraging its "translation" into technology other than direct public funding of academic scientists to commercialize

their work or of private enterprise; for example prizes or procurement contracts for novel drugs meeting independently assessed performance specifications, analogous to the way defense contractors are rewarded.

6. Rundle R, Large-scale shakeout looms for biotechnology firms, *WSJ*, April 6, 1994:B4.

7. On DNAX see Kornberg A, *The golden helix: inside biotech ventures* (Sausalito: University Science Books, 1995); Anon, Schering-Plough's prescription for success: pumping profits into R&D, *Business Week*, August 20, 1984:122 ("think tank"). As an example of a basic patent developed at DNAX, see Moore KW and Zaffaroni A, Hybrid DNA prepared binding composition, US Patent 4,642,334, issued February 10, 1987, on application of December 5, 1983; Rosa interview 1989 ("blue sky"); Feder B, Licenses for gene-splicing, *NYT* August 19, 1981:D1; Mowery DC, Nelson R, Sampat BN, and Ziedonis AA, The growth of patenting and licensing by U.S. universities: an assessment of the effects of the Bayh-Dole act of 1980, *Research Policy* 2001;30:99–119; Colaianni CA and Cook-Deegan RM, Columbia University's Axel patents: technology transfer and implications for the Bayh-Dole Act, *Milbank Quarterly* 2009;87:683–715. On Genenech's use of the Riggs and Itakura patent against competitors, and the later dispute with City of Hope, see City Of Hope National Medical Center v. Genentech, No. B161549, California Court of Appeal, Second District, Division Two, 20 Cal.Rptr.3d 234 (2004); Gullo K, Genentech must pay $300 million award, court rules, *Bloomburg Business News*, April 24, 2008, unpaginated news wire. Schering was willing to challenge Genentech over semisynthetic DNA method in law, but firms without deep pockets would not have been likely to.

8. Liebenau J, *Medical science and medical industry: the formation of the American pharmaceutical industry* (Baltimore: Johns Hopkins University Press, 1987); Dutfield G, *Intellectual property rights and the life science industries: past, present and future*, 2nd ed. (Singapore: World Scientific, 2009). It is worth noting that the "interpretive capture" (Dutfield, chap. 3) of the patent law by major drug firms that I am describing in the case of recombinant (and other) drugs in the US has also been attributed to the German patent law, so that patents on processes came largely to be interpreted as covering their products by the end of the nineteenth century; Gaudillière JP, Professional or industrial order? patents, biological drugs, and pharmaceutical capitalism in early twentieth-century Germany, *History and Technology* 2008;24:107–33. On the shifting terrain of nonobviousness in the era, see Sherman B, Patent Law in a time of change: non-obviousness and biotechnology, *Oxford J Legal Stud* 1990;10:278–87.

9. Pisano G, *Science business: the promise, the reality, and the future of biotech* (Boston: Harvard Business School Press, 2006), chap. 6 et passim; Hopkins MM, Martin PA, Nightingale P, et al., The myth of the biotech revolution: an assessment of technological, clinical and organisational change, *Research Policy* 2007;36:566–89.

Ampicillin: An antibiotic that is often used for *selection* of *transformants* when the gene conferring antibiotic resistance is carried by a *plasmid* or other *vector*. The ampicillin resistance gene encodes a protein that also breaks down related (beta-lactam) antibiotics of the penicillin family.

Band: A position on an *electrophoresis gel* (or other system to separate large molecules by size) containing nucleic acid or protein molecules of a particular mobility, usually visible through staining or radioactive tracers.

cDNA: Complementary DNA, so called because it is a DNA whose sequence is complementary to messenger RNA, produced by the virus enzyme reverse transcriptase on an mRNA template. The sequence of a cDNA generally reflects the final version of an mRNA actually used by the cell to *translate* into protein, after removal of intervening sequences in the primary transcript and other processing found in higher organisms.

Clone: A particular DNA sequence, or engineered organism bearing that sequence, removed from its original context and propagated identically. Cloning (molecular cloning) is the procedure of isolating a particular DNA sequence and moving it into a living organism such as a *transformed* bacterium so that it can be reproduced at will and manipulated.

Coding sequence: Those parts of a gene's DNA sequence that specify the synthesis of the particular series of amino acids constituting a protein, and that therefore correspond to the amino acid sequence of the specified protein.

Codon: A sequence of three nucleotide bases in a messenger RNA specifying the addition of a particular amino acid to a growing protein by the protein synthesis apparatus of a cell. The same sequence is found on one of the two strands in the *transcribed* gene.

Column chromatography: A class of methods used in biochemistry for separating diverse molecules in solution, where the solution is passed through a tube packed with a solid material that detains and therefore separates substances depending on their chemical affinity, size of molecule, or both as they move through it.

Denature: To cause, through heating or altered chemical conditions, nucleic acid and protein molecules to lose their stable three-dimensional structures so that they behave as simple strings.

Density gradient: A solution that varies consistently in density, generally produced in an ultracentrifuge tube with thick solutions either of sucrose or cesium chloride. Under equilibrium conditions dissolved molecules collect in bands of uniform density that can be separately removed, for example to isolate plasmids from chromosomal DNA. See also *sucrose gradient*.

DHFR: Dihydrofolate reductase is an enzyme found in bacteria and animals that, at high levels, breaks down certain chemotherapy poisons such as methotrexate. Placing a desired gene in close proximity to a DHFR gene on a vector, and culturing animal cells *transformed* with that vector in the presence of methotrexate results in *selection* of cells expressing the desired gene together with DHFR at high levels.

Electrophoresis. See Gel.

Exonuclease: An enzyme that breaks down nucleic acid strands by removing bases sequentially from the ends.

Expression, gene expression: The biological process in which a particular gene is transcribed and translated such that organisms manifest the trait it confers.

Fraction, fractionation: A systematically collected portion of a mixture or solution that has been treated by *chromatography*, production of a *density gradient*, or another method to separate different molecules within it.

Gel, gel electrophoresis: A wet polymer matrix, consisting usually of the plastic acrylamide or the sugar agarose, placed in an electric field and used to separate DNA, RNA, or proteins based on their size and/or charge. Typically samples are *denatured* under conditions such that their passage through the gel matrix, driven by attraction to the positive electrode, depends entirely on length, and produces a *band* of similar-sized macromolecules derived from the original sample. The solution in which a gel is "run" is called a buffer and may contain *denaturing* agents such as formamide.

Hybridization: The process in which a DNA or RNA molecule, or nucleotide sequences within it, will find and reversibly pair with complementary sequences in another DNA or RNA molecule. Hybridization conditions can be varied to produce different degrees of stringency, that is, degrees to which exact matching of sequence is required.

Library: A large collection of *clones* in which one might search or *screen* for a desired gene sequence. A genomic library is a collection of clones containing every piece of the chromosomes in a given organism.

Ligase: An enzyme that links together or *ligates* two DNA strands end-to-end.

Message, mRNA: Single stranded RNA molecules complementary to gene's sequence, containing both the coding sequence corresponding to the gene's end product and also sequences before and after it, necessary for the initiation and termination of protein synthesis.

Oligo, oligomer, oligonucleotide: A short string of nucleotides, generally synthetic DNA, that may sometimes be used as a *probe*.

Plasmid: A loop of DNA that occurs naturally in bacteria as an extra or satellite chromosome, often carrying genes conferring antibiotic resistance and capable of quasi-sexual transmission between bacterial cells. In genetic engineering, plasmids are used as a *vector* for *cloning*.

Polymerase: An enzyme that extends a polymer such as DNA or RNA by adding nucleotide bases, often to an existing strand or *primer* and according to the sequence of a complementary *template* strand. In *transcription*, mRNA complementary to the DNA sequence of a gene is initiated without a primer by an RNA polymerase.

Primer: A segment of single stranded DNA (or RNA) to which new nucleotides are added by a polymerase, extending the strand.

Probe: DNA or RNA, which may be single- or double-stranded, that is labelled with radioactive isotopes (or more recently, chemically) so that when *hybridized* with a diverse collection of nucleic acids it will reversibly bind to and thus identify complementary sequences of interest.

Promoter: A specific DNA sequence near or within a gene to which RNA *polymerase* will bind and initiate *transcription* to create an *mRNA*.

Regulatory gene: A portion of a chromosome that affects the expression of genes at some distance from it, often through production of a protein that interacts with the *promoters* of the regulated genes. Its biological function is indirectly to regulate other genes.

Restriction enzymes, restriction endonuclease: A type of enzyme that cuts DNA upon binding a specific sequence, and in the case of Type II restriction endonucleases, cuts in or immediately adjacent to that sequence. Some recognize and cut both single and double strand DNA (single strand cutters), while others require a double stranded recognition sequence.

Screen: A procedure, or principle on which a procedure is based, to identify a desired DNA sequence among a collection of *clones* or *transformants*.

Selection: A procedure to help identify desired cells, such as transformants, by eliminating all those lacking a particular genetic constitution. Frequently, transformants are separated from non-transformants in cloning by selecting those carrying a vector bearing an antibiotic resistance gene through culture in a medium containing the antibiotic.

Serine protease: A class of enzymes that digest proteins at specific sites, such as trypsin. The amino acid serine forms part of the enzyme's active site.

Shotgun cloning: The procedure of introducing random pieces of one organism's chromosomes into bacteria or other living cells. The alternative is to isolate the specific gene of interest, or pieces derived from it or its cDNA, before *transformation*.

Structural gene: A gene whose biological function is served directly by its own product (as opposed to *regulatory genes*).

Sucrose gradient: A method for separating nucleic acids by size in which they are made to pass through a dense solution of the sugar sucrose under high speed centrifugation, such that smaller molecules travel farther. The separate fractions are collected from the bottom, and each contains DNA or RNA of a certain size range. Largely superceded by gel electrophoresis.

Template: The DNA or RNA strand that specifies the sequence of a newly produced complementary strand, generally produced by interaction with a *polymerase* enzyme.

Tetracycline (Tet): An antibiotic often used for *selection* of *transformants* when the gene conferring tetracycline resistance is carried by a *plasmid* or other *vector*.

Transcription: The biological process by which sequences in a cell's active genes interact with an RNA *polymerase* so to produce complementary RNA, typically *mRNA* specifying a protein.

Transformant: Living cells that have been genetically altered or *transformed* by uptake of an introduced DNA molecule such as a *vector*. (Animal cells said to be *transfected*, to distinguish deliberate genetic manipulation from the usually spontaneous changes leading to conversion to a cancerous state, also called *transformation*.)

Translation: The biological process by which a cell's protein synthesis apparatus interacts with an *mRNA* so as to produce protein molecules specified by its sequence of *codons*.

Tryptic fragment: A specific piece of a protein produced through its digestion by the enzyme trypsin (see *serine protease*).

Vector: A segment of DNA such as a plasmid or virus chromosome that, after manipulation of its sequence and insertion of genes in the test tube, can be introduced into living cells so as to *transform* them.